Dedication

To my mother, Marilyn France, who taught me the power of poetry and the essentialness of empathy. To little Mina who found her namesake herein as a Sumerian unit of wealth. To the ancient Mesopotamian scribes and their diligent contemporary epigraphers whose timeless words are contained in these pages. To the more than fifty international journalists killed while trying to record the truth during the latest war in Iraq. And to "Nidaba" to whom we owe so much…

Establisher of the standard version thereof
was Nidaba [Goddess of writing and
literature],
she spun, as it were,
a web out of those words,
and writing them down on a tablet
she laid them ready to hand.

—Hymn to Kesh

Contents

Figure 1

Figure 2

Figure 3

Figure 4

Figure 5

Figure 6

Past-Present Ethno-Archaeological Linkages in Iraq's Marshlands

Edward L. Ochsenschlager

The Harvard Conference on the Iraqi Marshlands organized by Robert France in 2004 brought the serious situation in Iraq's marshes to public attention, pointed out the technical difficulties of restoring the marshlands, and initiated planning for ways to sidestep development obstacles, foreseen and unforeseen. Dr. France envisages several publications derived from the conference which will deal with the technical issues of attempting ecological restoration there. But in the present book, his focus is clearly on the people whose lives were destroyed by the drainage of the marshes.

For me, as an ethno-archaeologist, the relationship between the past and the present is of special importance. Without some connection between them there is little relevance to ethno-archaeological research, beyond attribution to that amorphous "human nature" whose boundary with human culture has never been properly determined. Luckily here in the marshes a connection between the modern marsh Arabs and the ancient Sumerians clearly exists. They are peoples who lived in very similar ecological settings, a similarity long suspected because of the nature and composition of ancient artifacts, and now verified in the recent thesis of Jennifer Pournell.

Hearing the Present Through the Voices of the Past

The ancient passages which Robert France chose for inclusion in this volume point out the importance of the general ecology, especially water, to the people of the ancient marshes. It captures their fear and despair when confronted with destruction of the marshlands on which they depended for their survival. In addition, these passages faithfully foreshadowed the modern devastation caused by the drainage of the great marshes. Here it is the past which illuminates the present. These words from antiquity help us to understand the agony and trauma of a modern people who, due to the circumstances of their isolation, have remained largely voiceless in our world. Through these expressions from the past we can at last understand the internal anguish of these silenced people of the present.

It was not only the silence of the sufferers that worked against our responding to their misery. This was not a sudden cataclysmic, headline grabbing, event; it was a development over time that swallowed up one small group of people after another. The waters slowly retreated, first here, then there, until the enormity of the total destruction finally dawned on the world media. Nor was there any single factor to which one could attribute this catastrophe at its beginning. There were increasing demands for water in the countries through which the Tigris and Euphrates rivers flowed. In Iraq there was a desire for dams that would contribute hydroelectric power to the nation and assure water supplies to growing cities. There was also the long-time attitude of many, including most western nations, that economic well-being could be enhanced by converting marshlands into irrigated agricultural fields. The crucial blow, however, was struck by Saddam Hussein, ostensibly to punish the Shiite people in the south who turned against him. Most people point to Iraq's defeat by the U. S. army in 1990 as the beginning of the tragedy, but in the area where I worked the Mi'dan people had been eliminated and the marshes preternaturally shrunk a year before the war commenced.

Even though the voices of the people of the marshes went unheard at the time of the disaster, the marsh Arabs found support in the voices and efforts of others. Amongst them are the contributors to this volume who are concerned for the peoples' present welfare, not consumed with nostalgia for their romantic past. Here Nik Wheeler has created a down-to-earth word picture of the Marshes in the mid-seventies that conjures up a vision of how things were on the eve of disaster, a prose version of his masterful photographs in the works he published with Gavin Young. The Baroness Emma Nicholson is filled with compassion for the people of the marshes but from the beginning exhibits a steely resolve to rectify the injustice she has seen and to restore the tranquility and dignity of the marsh people. Her tireless resolve to bring this inequity to public attention is well known. As a Member of the European Parliament, and through her work as chairman of the AMAR International Charitable Foundation as well as her immense contributions as World Health Organization Envoy for Health, Peace and Development, she has made immense contributions to the plans for marshland restoration and has brought urgently needed care to the displaced people of the marshes. Rasheed B. Al-Khayoun makes it very clear in his contribution to this book that the future of the marshes lies in the restoration of a newly franchised population, not as some kind of Disney World exhibit, but as a modern society with modern lives and modern goals informing their traditional society.

These essays provide hope for the future and bring a sense of dedication to the renewal of the environment as well as a new beginning for a traumatized population. They also bring back for me haunting memories of what a fully integrated life was like in the marshes where the village rituals of birth, marriage and death, and the centrality of crafts and occupations in daily life formed a framework for village traditions, morality, and the sense of belonging. Above all, I am haunted by memories of the children who went about their work or play with so much energy and speed that they looked like chaff caught in the wind on the thrashing floor. It is these children rooted in old traditions who provide hope for the future of their people. They must provide a new framework for meaningful survival and personal contentment as they return to their former homes.

Seeing the Past in the Present

Just as the past can help us understand the present, so can the present help us understand the past. Dr. France has included in this volume 31 photographs taken during the Field Museum of Chicago's 1934 expedition to southern Iraq. They were later donated to Harvard University by the leader of the Expedition, Henry Field, and are now housed there in the Peabody Museum of Archaeology and Ethnology. These pictures, when combined with photographs taken by John Henry Haynes in the late 19th century in conjunction with the University of Pennsylvania Museum of Archaeology and Anthropology's excavations at Nippur and now in the Museum's Archives and also recently published in my book about the marsh dwellers, provide us with a view of life in the marshes at the end of the 19th and the beginning of the 20th centuries. Added to the photographs of Gavin Maxwell, Wilfred Thesiger, Gavin Young and Nick Wheeler along with my own, we now have a published panoramic view of life in the marshes which covers nearly a hundred years. On the one hand these photographs illustrate a way of life which has vanished; on the other, in conjunction with the ethno-archaeological study I present in my book *Iraq's Marsh Arabs in the Garden of Eden*, they give us valuable insights into what life must have been like thousands of years earlier.

Some ethnographic parallels drawn from this information offers direct evidence for the preparation of raw materials and the manufacturing of artifacts in the ancient past. Sometimes we can restore complete processes that have left no trace in the archaeological record and even infer the nature and composition of missing artifacts from artifacts that are present by supplementing archaeological evidence with ethnographic knowledge. Furthermore, knowing the worth of raw materials and the time and skill involved in the manufacturing process gives us a way to determine an artifact's value to the people who made it in ancient times. Other ethno-archaeological discoveries lead us to question previous archaeological conclusions or offer interpretive options formerly overlooked or unrecognized. Still others help us unlock the meaning of archaeological phenomena that may have no direct parallels in our previous experience.

And it is in this spirit of temporal commingling that *Wetlands of Mass Destruction: Ancient Presage for Contemporary Ecocide in Southern Iraq*, is truly a book of many facets. For herein Robert France gives us not only a new perspective on the past and on the present, but also provides a great hope for the future.

Preface

Out of Time's Abyss:
Voices of Portent

I would tell you about it,
would bemoan the bitterness
of my existence.

—Lamentation Poem

Mine heart within me is broken...
the land mourneth;
The pleasant places of the
wilderness are dried up.

—Jeremiah

For the religious minded in the modern Western world, the Bible—even the Old Testament—is often approached and interpreted as a source of guidance for contemporary affairs. The devoted, therefore, wouldn't think that there was anything at all strange about reading a passage that might have been written over two thousand years ago and finding not only revelation but also relevance for their present day lives. The Old Testament is rife with prophecy throughout its pages. The lesson of Daniel and King Belshazzar is, after all, that God's word is veritably "written on the wall."

Also, for contemporary adherents of the "New Age"—a movement that very much contains at its core a recycling of age-old truths wrapped up in flowery, modern-sounding prose—the voices of the past are often reworked and reconfigured in order to show prescience for today's events. In light of this, there has been a minor growth industry in interpreting the prophecies or meditations of such sages as, for example, Nostradamus or Hildegard von Bingen. Nowhere, however, does the expression "I told you so" ring more true than in ancient Mesopotamian (specifically Sumerian, Akkadian, Babylonian, Assyrian) literature. It is in Mesopotamia—literally the "land between the rivers" Tigris and Euphrates—where it all really began.

Many years ago the noted Sumerian scholar S. N. Kramer wrote a book entitled *History Begins at Sumer* in which he examined over two dozen "firsts" that had their roots in that ancient civilization of four thousand years ago. Of all the many and marvelous inventions of the Sumerians, it is often asserted that their greatest contribution to human cultural evolution was the invention of writing. The scribbling down of words upon clay tablets with reed styluses by these earliest of scribes has left a record of communication that continues to inspire the imagination of historians. Such writings, however, were not always limited to the pervue of scholars. Indeed, in the last decades of the nineteenth century, when the Western world was abuzz with all things Mesopotamian due to the new archaeological excavations there, the general public was kept captivated by reading about the latest translations of the tablets on the front pages of their newspapers. Part of this fascination arose from the anticipation that each new translation might somehow reveal or "prove" a Biblical story.

This worldwide Mesopotamian fever reached its peak when the translated passages of the Gilgamesh epic—in particular, those dealing with the Flood myth and its Noah-like protagonist—were brought to light. Here, some Biblical believers found all the "evidence" they needed to battle the evolutionists and geologists who were beginning to capture much of the public attention with their own very different (and modern) interpretations of global history. Armed with these powerful voices from the ancient past, the religious-minded defended their belief in a divinely-inspired deluge in the land of their ancestor Abraham.

Today it appears that these selfsame voices from the abyss of time can offer an interesting vantage point from which to examine the bleak environmental and human rights situation that has recently unfolded in the southern Iraqi marshlands. For example, the United Nations Environment Program's 2001 report *The Mesopotamian Marshlands: Demise of an Ecosystem*, begins with the following quotation from Gilgamesh:

> *Ever the river has risen and brought us the flood,*
> *the mayfly floating on the water.*
> *On the surface of the sun its countenance gazes,*
> *then all of a sudden nothing is there.*

In 2004 I organized an international, multidisciplinary conference at Harvard University titled "Mesopotamian Marshes and Modern Development: Practical Approaches for Sustaining Restored Ecological and Cultural Landscapes" (the present book contains excerpts from three of the presentations at the conference). As conference preparation got underway and I delved deeper and deeper into the literature about all things past and present concerning Iraq and Mesopotamia, I was often struck by the seemingly prophetic nature of much of the historic material I was reading. It became impossible to escape a feeling that many of today's events were what might be called "déjà vu all over again."

Soon my little notebook of jotted-down quotations cribbed here and there from the literature began to fill up and thus call for more thorough attention. Eventually what had begun as an ancillary hobby tangential to my main purpose of researching the landscape history of Mesopotamia, grew into an investigative project in its own right. With this new purpose in mind, I honed my "search image" for Mesopotamian inscriptions of ancient presage for the contemporary ecocide in Iraq that was to be the subject of the upcoming conference and related professional books. The result was that the nearly fifty texts I consulted yielded over two hundred inscriptions appropriate to the intended purpose. Also, I augmented the Mesopotamian inscriptions with quotations found by pouring through various Old Testament books that dealt primarily with events transpiring within Babylonia and Assyria at the time of the Exile.

In the following pages, I have organized the inscriptions and quotations into three major categories: *Water* ("A Swamp He Made Into Dry Land"), *Ecology* ("The Lord's Work Kills the Marsh") and *People* ("A Reed Pipe of Dirges"). Within each major category, I have arranged the material into either five or six sub-groups; the first two concern Saddam and the war against the south, the next two or three concern the effects of those actions upon various attributes of water, ecology, or people, and the last one concerns the emotional toil of these effects upon the surviving people.

What will almost immediately strike the reader familiar with the massive environmental and cultural destruction in southern Iraq (see Introduction) is the incredible prescience of many of the ancient inscriptions. At times what is written is almost eerily accurate in its prediction of contemporary events. Indeed, some of the writings could almost be used verbatim as modern-day newspaper headlines: "The men knew not water," on river diversions and marsh drainage, "Not even a reed marsh was to be seen," for the destruction and disappearance of the wetlands, "The mantles of radiance were lost," on the consequent loss of biodiversity, and "He plotted evil, to devastate the land, to destroy the people," for the deliberate ecocidal and genocidal intentions of Saddam himself.

Of course, in their original context, these inscriptions had meanings often quite different from the ones which I've associated with them here. Although the ancient Mesopotamians resided in what appeared to be a verdant land of health and promise, their fortunes were still dependent upon the vagaries of climate and its implications on water supply and agricultural bounty. In addition to this precarious existence based on sustenance agriculture, the common citizens were also at the mercy of despotic rulers who seemed to take great pains at every opportunity to engage each other in an endless cycle of war. No wonder then, that when things took a turn for the worse, either through the imagined acts of various gods upon the climate or through the realized actions of cruel warlords upon the landscape and people, the writings of the scribes expressed a profound sorrow at the miserable circumstances of their lives.

One cannot help but try to imagine what the writings of present-day marsh dwellers in southern Iraq might say about their abysmal situation. Until these witness testimonials are recorded for prosperity and for us to learn from, it is not unjustified to turn back the pages of time to what their early ancestors might have said. Presage is, in the end, a gift—in this case, a gift from the loquacious voices of the past to the cruelly silenced voices of the present.

The blessing of portent is that with foreknowledge one can act to protect the future. The devastation of the marshes of southern Iraq and the consequent loss of a unique piece of cultural heritage extending back in time to the Sumerians is indeed a tragedy. But alas, it is not a tragedy that has not occurred elsewhere around the globe in the past, nor is it one that we can be assured will never occur again somewhere in the future. We can learn from the power of words and, if wise, we can heed the mistakes of the past. If, as is contended, history really did begin in ancient Sumer, then we should not be surprised that the powerful words echoing out from the abyss of time can provide a canticle of warning to help guide us along a path respectful of both rarified nature and indigenous culture in today's troubled world.

Prologue of Yesterday

In the Garden of Earthly Delights:
Eden Found

With the river filled with flowing waters,
the marshes stocked with life...
Abutting heaven.
Its awe and glory
cast upon the country...
Grown up 'twixt heaven and earth.

—The Cylinders of Gudea

Let the waters bring forth abundantly
the moving creatures that have life,
and fowl that may fly above the earth
in the open firmament of heaven.

—Genesis

Ecologists refer to the boundaries between different types of ecosystems as "ecotones." Usually these boundaries are blurry distinctions between, for example, mountain and valley biomes in alpine regions, or between freshwater and saltwater mixing zones in estuaries. In desert regions, however, where the influence of water is observed so overtly, ecotones assume a much sharper edge. The contrast here between areas of stark barrenness and those of abundant life can be quite dramatic. It is no wonder then that these landscape typologies have entered our consciousness in the form of those two defining archetypes through which we dichotomize our natural world: "wilderness," and the "garden."

The Garden of Eden is possibly humankind's oldest legend. The word "garden" has its origin in old Hebrew in which "gan" was a specific type of walled enclosure (with the ancient Persian name for such being "paridaeza" from which we get the modern "paradise"). The etymological distinction between garden and wilderness becomes a little conflated in the origin of the word "Eden," which is possibly from the ancient Sumerian-Akkadian "edinu" referring to steppe areas outside of the cultivated land. In old Hebrew, however, the word "aden" refers to a cool place for regeneration. In this respect, we can identify Eden as being a "gan'adan," meaning an enclosed, cool garden. By the Middle Ages, Eden came to be looked upon as the "garden of earthly delights," that well-watered place where God strolled in harmony with all His creations. And though a minor growth industry developed among debating theologians and cartographers as to where it was that the Garden of Eden had actually existed, most reasonable guesses situated it somewhere in the verdant floodplains between the rivers Tigris and Euphrates.

The marshes of southern Iraq once comprised one of the most ecologically important wetland ecosystems in the entire world. Before their draining, the marshes covered an area of up to twenty thousand square kilometers and were sustained by mountain snowmelt being carried through the bordering desert by flooding rivers. Although the precise location of the marshes has shifted considerably over time within this dynamic riverine system, until recently they shared the common trait of always being an incredibly diverse and abundant oasis for wildlife. It has been estimated that as many as three million birds from over one hundred waterfowl species used the marshes during their annual migrations to and from Africa and central Russia. The marshes also played an important role as the nursery for thousands of fishes and other aquatic fauna that moved back and forth between the vegetated habitat and the open waters of the Arabian/Persian Gulf. Other species, some endemic only to these marshes, included mammals such as otters and wild boars. Although such a rich biodiversity as that formerly found within the southern Iraqi marshes would have been significant anywhere in the world, it is the dramatic contrast between this productive region and the near desolation characteristic of the surrounding deserts that made the place seem so marvelous, unbelievable, or indeed to some, even divinely created. It is no wonder that humans have always been attracted to the area, first, physically as a bountiful place for residence, and then later, conceptually as God's original garden from which we all sprang.

Around the world, the history of indigenous culture is often the history of wetland inhabitation—no more so than within the marshes between the Tigris and Euphrates rivers. The first human marsh dwellers in the area of southern Iraq probably moved there from the Persian/Arabian Gulf as the later region became inundated following glacier melting at the end of the last Ice Age. Carrying their flood myths with them and into our Western consciousness, the dwellers of the original submerged marshes had probably first settled there after descending from the fertile crescent of the Persian plateau. Later, the Sumerians (of still mysterious origin) arrived and created the world's first civilization at the border of the marshes, sustained by the abundant wildlife and especially by harnessing the waters for year-round crop irrigation. As ancient Mesopotamia developed, Sumerians blended with Akkadians and Babylonians to influence the Assyrians. Another common characteristic of the marshes throughout this period (and even extending into that of the more recent Islamic Middle Ages) was that they served as an ideal refuge for those fleeing from the central authorities. As a result, the marshes became an ethnic melting pot of complex racial origins such that the word *Ma'dan*, or marsh Arabs of today, is really a convenient umbrella term for a variety of marsh dwellers living in and on the margins of this vast wetland.

Over the last century, a few western travelers were able to penetrate the inner sanctum of the marshes to observe and record the lifestyles of the residents, who proved seemingly little changed from the images of those earlier dwellers left behind four thousand years ago on various Mesopotamian carvings. Because there is an all too real and very easy danger of succumbing to overt romanticism when considering the modern Ma'dan, aided by the powerfully haunting prose of those fortunate to have visited them, it is almost impossible for many to resist the temptation of likening these marsh dwellers to the imagined original and innocent inhabitants of the Garden of Eden. From an objective environmental perspective, however, what can safely be stated is that until the recent destruction of the marshes, the human inhabitants there lived an ecologically sustainable existence such as is found only rarely elsewhere around the world today. Depending upon fishing and hunting within the marshes, and upon sustenance agriculture along the edges of the marshes, the lives of those in the marshlands were in true harmony with their environment. Their reliance upon the tall reeds of the marshes for constructing their artificial islands and homes, and their close partnership with their semi-domestic water buffalo— the latter sustained by collected reed fodder and then in turn sustaining humans by supplying milk and dung—seems almost too good to be true: a dream of symbiotic mutualism in a magically timeless landscape.

Unfortunately, in the last decades of the twentieth-century, that dream became a nightmare, and time ended for many marsh inhabitants, both human and nonhuman.

Introduction to Today

Wetlands of Mass Destruction:
Paradise Lost

May you speak it not out,
May you speak it not out:
Destruction!

—The Verdict of Enlil

Destruction upon destruction is cried;
for the whole land is spoiled.

—Jeremiah

In the early 1990s immediately following the first Gulf War, the United States encouraged a Shi'a uprising in southern Iraq with the aim of further destabilizing Saddam Hussein's government. Fears by many, however, including those in neighboring governments such as the Sunni-run Kuwait, led the United States to back down and fail to offer military support, thus dooming the nascent uprising. This paved the way for Saddam to vent his frustration at loosing the Gulf War by attacking a much more manageable foe: the dissidents in the south.

As had happened time and again in this region, the dissidents fled into the expansive marshes in attempt to hide and thus thwart the tyrant's vengeance. Using air power, Saddam's Ba'thists sought out and systematically attacked the fugitives, and in the process indiscriminately killed thousands of, in his mind, meddlesome marsh Arab collaborators, by napalming their homes and strafing the survivors. Those long-time inhabitants who survived the firing of their villages began to flee the marshes to escape the Ba'thist's wrath. Rape and torture were rampant. And it is here where Saddam's next move raised the bar on what in all eyes must surely count as being one of the most egregious acts of institutional evil of the last century.

With a propaganda campaign in Baghdad insisting that the Ma'dan were not really true Arabs to begin with, even going as far as to dehumanize them by likening them to monkeys, the stage was set for Saddam's pogrom. The way that he accomplished this amounts to one of history's worst "ecocides," that is, the purposeful destruction of an environment to bring about a deliberate genocide against a targeted people. In localized areas cattle were slaughtered, reed beds were burned, and fish were poisoned; and most severe of all, water was removed from widespread regions. Masquerading as a plan to drain or "reclaim" the marshes for increased agricultural production (as has been accomplished elsewhere around the world), Saddam's hydro-engineers began their environmental onslaught in earnest.

Drainage plans initially laid out by the British in the 1950s and perfected by the Ba'thists as early as the 1970s and 1980s (both based on an eroneous belief in the marshlands as being landscapes completely devoid of economic utility), combined with the already strangled water flow in the Euphrates and Tigris Rivers due to massive dams in Turkey, systematically destroyed the marshes—one of the most precious ecological jewels of the world—at a prodigious rate rarely previously achieved anywhere in the world. Although there is a rich history of the use of water as a weapon of war in ancient Mesopotamia, the enthusiasm coupled with the technological acumen of Saddam's engineers meant that the extent of destruction of the wetlands was profound. In little over half a decade, more than 90 percent of the marshes were either drained or starved of water due to construction of a comprehensive network of dikes, dams, ditches and diversions.

The effects of the massive destruction of the wetlands upon the residents were as might be expected. Biodiversity plummeted. Species capable of escaping like migrating birds, that could alter their flight patterns, were able to do so. Other species that could not flee, such as many mammals, fishes, reptiles, and amphibians, were extirpated. And accomplishing Saddam's "final solution" of ethnic cleansing, thousands of the Ma'dan, and also Bedouin who lived along the edges of the marshlands, either fled across the border into Iran to begin their decadal-long internment in the refugee camps there, or emigrated to Europe or America.

Today the marshes are but a vestigial remnant of their former glory. Recent photographs bring feelings of shock, awe, and overt sadness to even the most inured or jaded observers. Over vast areas, the dry sand is encrusted with a surface layer of salt due to evaporation of the last water. A few water buffaloes, now grazing upon small clumps of salt-tolerant plants where once they waded and fed freely upon abundant marsh reeds, are almost all that remain to signify the vibrant domestic life that once existed there. Villages that escaped direct burning, now lie mostly empty, forming curious mounds rising above what had once been the wetland plain. Almost all wildlife are either long dead or have moved elsewhere to those few patches of marsh that somehow escaped Saddam's ecocidal actions. The land, once the most fertile in the ancient Near East, has become a vacant, desolate wasteland where naught, either human or nonhuman, visits or resides.

The scale of environmental mass destruction of the marshes in southern Iraq is almost unparalleled, perhaps only being matched in recent years by the shrinkage of the Aral Sea. And added to this of course is the harrowing of the marsh Arabs to produce the world's most famous "environmental refugees," a shameful bookend to a century of genocide that had begun in

Armenia. The following album comprised of inscriptions and quotations originating thousands of years ago provides, at the same time, a frightening storyline to this most contemporary of environmental and human rights disasters.

Part One: Water

"A Swamp

He Made

Into

Dry

Land"

Water – Saddam

"What the evil
winds had
sent hither"

The weapon... loving lord...
put an end to the rivers,
he came and dried up
the waters like a strong wind...
That Baghdad boil—
a breaking out of which
on the nose is not pleasant—
that twisted tongue...
What the evil winds
    had sent hither.

The Lord piled up a dam at
the edge of the sea;
a swamp he made into dry land.

And I will make the rivers dry...
and I will make the land waste,
and all that is therein.

Revolution, chaos, and
    calamity
will occur in the country.
A dreadful man,
son of nobody,
whose name not mentioned,
    will arise.
As king he will seize the throne,
he will destroy...
the people of the land...
The marshes and the rivers
Will fill up with sand.

The king matched
Tigris water with
Euphrates water.

I shall... shift the course
of the irrigation channels
and canals.

He diverted water
into the boundary channel...
in the direction of the bank
of the Tigris.

All its waters...
    I shall dry up

And the land shall be
desolate and waste...
The river is mine,
and I have made it.

He dug the... canal and
extended it's
     far end to the sea.

Nowadays these water
do not any more
rise up in the earth...
What once was scattered
be gathered together,
What had been absorbed
     by swamps...
He gathered and
threw into the Tigris.

I will cut off the water
necessary to their lives.

Water—War

"Disturbed
was the
Tigris river"

Disturbed was
    the Tigris river,
it became constricted,
    roiled, stirred up.

Terror holds the exalted river,
    the Euphrates.

The Tigris
    is surrendered
to him, as to a
    rampant bull.

Cities are a desolation,
a dry land,
and a wilderness,
a land wherein
    no man dwelleth.

A day of doom,
the assault screamed
    wrathfully,
like a formable serpent
it hissed from among
    its people,
it wiped up the rivers...
it gashed the earth's body
made painful wounds.

Sometimes there is
hostility in the land.
Ever the river has risen
and brought us the flood,
the mayfly floating on the water.
On the face of the sun
its countenance gazes,
then all of a sudden
nothing is there.

Water - Desiccation

"The men...
knew not
water"

The men...knew not water.

The parched earth all around.

The irrigation water
was silent...
Even the canals
were quite silent.

Here is no water.

He was not even allowed
    to drink water.

My houses at the outskirts
    of the city
were verily ravaged before me...
In my city's river
    dust is gathered,
foxholes are verily made therein,
flowing waters are not
carried in them.

The waters of the ground
coming from below
did not flow
out over the fields.

A river that sparkles not
abounds not
in flowing water,
carries not water.

My delightful plain
where its delicacies
were prepared
is verily parched
like an oven.

Stolen waters...

Water—Pollution

"That water...has gone bad"

That water,
The surface of it
    has gone bad.
How could you drink
    that water?

Its central plain...
grew reeds of lament...
the water flowing sweet
flowed now as saline waters.

Instead of your
sweet-flowing water,
    bitter water will flow.

Water - Laments

"I shall
weep for you,
the pure
Euphrates"

I shall weep for you,
the pure Euphrates,
With whose water
in waterskins we used to
refresh ourselves.

The Tigris river
did not grandly rise up...
it carried not fresh waters,
so men could not
dip water pails
at the quay.

They came to the pits,
and found no water;
They returned
with their vessel empty...
because the ground is chapt.

Water I have not drunk,
    tears were my drink.

Even the control of
heaven and earth was undone,
the springs diminished,
the flood-water receded.
I went back,
and looked and looked;
it was very grievous.

He who travels out
on a path with water
shall return along
a way of dust storms.

By the rivers of Babylon-
There we sat down and
there we wept.

Figure 7

Figure 8

Figure 9

Figure 10

Essay One

Witness to a Lost Landscape:
The Marshes in the Mid-Seventies

Nik Wheeler

I have been fortunate to have met some of the key players who have been putting in so much time and effort into the very worthwhile project of restoring the marshlands to at least some their partial former glory. My first trip to the marshes was in late 1974. At that time, I was living in Beirut and received an assignment from *National Geographic* to shoot a story on the marsh Arabs and their life in that area. The writer of the piece was a British journalist, the correspondent for the *London Observer*, Gavin Young. He had been in the marshes several times, years previously, accompanying the famous explorer Wilfred Thesiger. So Gavin had very many good contacts in the area which turned out to be absolutely crucial because the logistics for traveling in the marshes in those days was quite horrendous: there were no hotels, no restaurants, no stores even, no commercial transportation, and no water taxis. So we had to arrange everything privately through Gavin's connections. Fortunately he was able to organize this through a local holy man the use of a *tarada* or war canoe and together with four marsh Arab canoeists who paddled us from village to village.

We also had a minder from the Ministry of Information because any time foreign journalists wanted to travel anywhere in Iraq, not just in the marshes, someone from the ministry had to go along to make sure we were not talking to anyone we shouldn't have been and that we kept strictly to the assignment we had been given permission to cover. This was the period when President Al-Bakr was the leader and Saddam Hussein the Vice-President but in reality, the power behind the throne. Teriq Aziz was the Minister of Information and the one instrumental in giving us permission to travel in the marshes.

I have said this before—and this is very important—even back in those early days I do feel that the Iraqi administration was thinking, and had back in their minds, the intention of developing and draining the marshes. And so they felt that our trip there would be the last documentation of the way of life there before they "helped" it on its way into oblivion. They gave us unprecedented access to the region and this was quite unusual because up until the time of our trip, there had probably only been three or four Westerners who had spent any substantial amount of time deep within the marshes. And even the Iraqi themselves rarely ever set foot in the marshes;

it had always been considered a haven for draft dodgers, for political malcontents, for people on the run, and for common criminals, etc. So it was really a forbidding place for most Iraqis who were used to spending their time on dry land. The marshes were a whole other world.

In addition to giving us access and a minder from the ministry, we also had a security man from the town of Amara and we were also allowed to bring in a double-barreled shotgun and a box of cartridges. This was so we could hunt some coot and wild duck which were plentiful in the marshes and this allowed us to eat something other than the rather bony fish which were the other staple in the marshes.

For the first trip, we were there for ten days in the winter months, and after the *National Geographic* piece was published, we were asked to go back by a British publisher and so the following spring—which was a very different time of the year in terms of the marsh climate (it was much warmer, the days were hazier)—we went back for another ten days to take more pictures for the book. And in that period I was given access to a Soviet-made helicopter which was based in Basra and was therefore able to fly over the marshes for a couple of hours shooting unlimited photos. This was quite an unusual step for a very paranoid and oppressive government to allow a foreigner with a camera that was always considered a suspicious object (you could often be arrested in Baghdad for taking photos of a bridge or a school because these were considered to be possible military targets). So suddenly to be up in a helicopter without anyone telling you not to use your camera, was quite a wonderful feeling and a great help.

My photographs from those trips can be broken down into a few categories: general geographic overview, historical museum artifacts, portraits showing the different physiognomies of the peoples who lived in the marshes at the time, their very special houses and how they were constructed, boats and fishing (a very important part of their livelihood), daily life, domestic animals and wildlife, and finally modernization and the portent of things to come which ominously heralded what did transpire several decades later.

From a technical point of view, in those days I was using an old Nikon-F that had no light meter, and thus had to rely upon a hand-held meter to take exposures (state-of-the-art technology at the time). And for the second trip I had a Nikomate-EL that had a little arrow between brackets to inform you of the correct exposure. The film was predominantly Kodachrome 25, which was a wonderful film, and was augmented by some Ektachrome 64 that was pretty awful, leaving a green cast to several of the photographs.

The marsh Arabs of the 1970s lived lives similar to those of their ancient Sumerian predecessors. Cylinder seals exist which show Sumerians paddling through the marshes. Unfortunately during the recent looting of the Baghdad Museum, the entire collection of seals was stolen and this particular one was not recovered when some of the more famous artifacts were returned. There also used to be another relief from the Baghdad Museum showing the same sort of dwelling used five thousand years later by the marsh Arabs. Reliefs from the British Museum show the forces of King Sennacherib entering the marshes in reed boats as they searched for a southern army he was pursuing. As well, there are alabaster busts from the Basra Museum of ancient Sumerians who closely resemble modern-era marshland inhabitants. And there is two-foot silver model of a boat found at ancient Ur that seems to be an exact copy of those used in the marshes until the 1970s.

Typical villages in the center of the marsh were composed of groups of little platforms, each inhabited by a single family and crowned by a single reed house. These platforms over the years had been made from reeds and mud brought in from the dryland and tamped down with more rushes and earth added until they have basically a small island surrounded entirely by water. Scenes of these villages with the giant bed of reeds stretching off in the distance remain powerful memories; as do images of children in boats, who from the age of six were allowed to paddle around in canoes and spend most of their young lives out on the water.

There were really little indication of religion in the marshes that I observed; no mosques of any kind; only occasionally were people seen praying; and when someone died they would be transported to the nearby holy towns for burial. But I was not there during Ramadan so can't really comment on how seriously it was observed. But certainly I saw very little visible signs of the presence of Islam.

I strongly remember one character: the marsh Arab clown who in the evening would come in from the reeds and start rustling the side of the *mudhif* or the communal gathering place, resulting in all the kids screaming. He would put flour on his face and wrap himself in a black garment, pretending to be the ghost of the marshes.

Mostly women would go off and cut reeds and paddle them back to the village. The green reeds were used for water buffalo fodder, the giant yellowish reeds used for house construction. For the latter, the reeds would be gathered into pillars on either side of the location of where the house would stand following the surface being stamped and raised. The bases would then be bound using only reeds. Some of the communal mudhifs had fancy and beautiful lattice siding. After the mats were placed on top as roofing material, they could be raised, thereby allowing for a current of air to come through for cross ventilation in the summer when it could get quite very hot (above 120° with maximum humidity). Some of the mudhifs were up to a 100 feet in length, and about 16 feet high by 20 feet wide.

At night, men would come into the mudhif and sit around and gossip about the affairs of the day. The women had an area at the back of the main hall where they would prepare the food and, unless they were very old, would rarely come in to sit with the men.

There were motor launches from the main island towns to the edge of the marshes where individuals could pick up a bus or a taxi if they needed to go the nearby towns for markets or other reasons. And often people would hitch their canoes to the large boats so they could return home on their own whim. These private boats were made from mulberry and other types of wood imported from Indonesia and constructed with only a hammer, an adze, and a few nails. The exterior of boats were covered with bitumen from a nearby town which had been rolled onto the outside surfaces for waterproofing.

Most of the fisherman preferred to use long, 5-pronged spears in the shallow water of the marshes. They would also look down on others who instead would use nets for fishing, which was in those days was considered not very sporting. The fish were very bony and carp-like with muddy tasting flesh that one soon became, at least for a Western palate, quite tired of eating. The spears were made by the Mandeans, an ancient sect worshing John the Baptist, and whom were famed for their metal work.

Inside the houses there were slow-burning buffalo dung fires and large cooking pots. Unleavened bread was baked atop a large stone. There was always an outside kitchen for the washing up. Breakfast was usually an omelet with pancakes made from rice flour. Banquets consisted of rice, chicken, fish and bread, all washed down with some good marsh water. While out in the marshes in their boats, people would ruffle the surface of the water to displace floating particles and simply flick the liquid up into their mouths. There was no other source of water at all, and though I personally had no difficulties, I do understand that those who lived there continually did have health problems. When back at their island homes, often very strong Bedouin-type coffee would be drunk in small porcelain cups.

For wedding ceremonies, the bridegroom's party would crowd onto a small motor launch and head off to pick up the bride at her house, their rifles waving in the air. The fathers of the betrothed would seal the deal with a handshake.

I was quite amazed me when I recently visited the Peabody Museum at Harvard University and saw the photographs from 1934 Field Expedition to the marshes which seemed almost identical images to many of those I took forty years later. For example, photographs of subjects such as a woman grinding rice seem almost interchangeable, and what really struck me therefore was how little the marsh life had changed over all that time.

One of the few signs of the industrial revolution reaching the marshes that I saw was an old dilapidated diesel engine used to power a mill for grinding rice into flour to be stored in sacks. Another major source of income were reed mats which the women would pound flat after the men had split the reeds. The mats, composed of reed stalks woven together, would be rolled up and transported by canoe to the nearest dryland place where all along the channel everyone would deposit them, and from where they could be transported by truck to the nearest town or even as far away as Baghdad for sale.

Water buffalo played a major role in the life of the marshes. Women would take the buffalo into their watery "fields" to graze on the reeds. Boys would bring back reeds as fodder for the buffalo. The buffalo were kept mostly for milk and fuel. The dung was made into paddies mixed with straw and left out to dry, often on the roofs of buildings. And given their immense value, when the buffalo were sick they would be tended like cherished family members.

I was fortunate to have seen the marsh wild boar which could reach up to four-feet to the shoulder, weigh three hundred pounds, and thus be very dangerous. Pelicans and storks filled the skies. In the old days, the marsh Arabs would use pelican beaks for drums. The most dangerous animal of all, however, was the dog. Every little island had two or three, and you had to be very very careful if you set foot at someone's house because you would likely be attacked by these vicious animals. So you had to make sure that you had a stick handy or that the owner knew you were expected.

I want to conclude by stating that even in the 1970s the marshes were slowly being dried up. Dikes were being built, no doubt based on the British plans from the 1950s for providing more arable land. Once from the air, I saw and photographed a newly constructed dike with dry land on one side while the other was still covered with abundant water. Several villages even then showed the signs of being drained so substantially that their many islands, once separated by surrounding water, had now become joined together as a result of the desiccation. Indeed, I flew over one village where homes could be seen which were completely isolated from all water, the surrounding marsh having been totally dried up. And at another village, the flares from oil wells could be seen off in the distance, a sign of what was to come. So the point I would like to make is that although Saddam Hussein was the one to later execute the drainage plan in its formal finality during the 1990s, the destruction of the marshes had really been in the works for many years before that. In this respect, Wilfred Thesiger was unfortunately quite prescient when he wrote in the mid-Sixties that "Recent political upheavals in Iraq have closed this area to visitors. Soon the Marshes will probably be drained; when this happens, a way of life that has lasted for thousands of years will disappear."

Part Two: Ecology

"The Lord's

Word

Kills the

Marsh"

Ecology - Saddam

"I shall
dig out
reed
thickets"

I shall finish off the land
    and count it as ruins...
I shall devastate
    the cities
    and make of them
    a wilderness...
I shall dig out reed-thickets
    and graves and
I shall burn them...
I shall leave no life.

I will make the land desolate.

The lord's word
    kills the marsh.

I will destroy also
all the beasts thereof
from beside
the great waters.

You have destroyed
    the indestructible,
you have made perish
    the imperishable.

The date palms
which were
the growth
of the country,
I destroyed.

So that the fishes...
and the fowls of the heaven,
and the beasts of the field,
and all creeping things
that creep upon the earth...
shall shake at my presence.

The terrible,
the unsparing
King,
whose onslaught
is a cyclone,
who overwhelms
the land.

I reduced the country...to
a rubbish-heap,
as if tossed together
by a hurricane.

The earth
    is devastate,
        in my
            supremacy.

He took and devastated
all the lands...
To dust and ruins he made all,
leaving not enough
for birds to rest on.

The ruler rains destruction
upon the hostile regions,
rendering them
ruined and desolate.

He has surely made an end
of what is of the very life's breath...
in the marsh.

Bring evil from the north,
and a great destruction.
The lion is come up from his thicket,
and the destroyer...
is on his way;
he is gone from his place
to make thy land desolate.

Ecology—War

"The
earth's
outcry"

The earth's outcry...
The warlike... man...
Opened up
   a hole
      in the earth.

Earth,
a mortification!

Before the warrior went a huge
irresistible tempest,
It was tearing up the dust,
depositing it
evening out hill and dale,
filling in hollows,
live coals it rained down,
fire burned, flames scorched,
tall trees it toppled from their roots,
denuding the forests.
Earth wrung her hands against the heart,
emitting cries of pain.

A horrible thing
is committed
in the land.

The whole land trembled
at the sound of the
neighing of the
    strong ones;
for they are come,
and have devoured the
    land,
and all that is in it.

How do the beasts groan...
for the fire hath devoured
the pastures of the wilderness.

Heaven's wild...
   who have trampled...
      under your hoofs.

The land perisheth
and is burned up
like a wilderness,
that none passeth
through.

How long shall
the land mourn,
and the herbs
of every field
whither?

The flying birds
had their heads beaten in,
their wings trailed the ground;
the fishes down in the deep
the war storm smote,
they were gaping in death
the gazelles and wild asses...
in its path were famished
the landscape was burnt off
as if denuded by locusts.

The whole land
is made desolate...
no flesh
shall have peace.

The reeds
they are burned
with fire...
and the men
are
affrighted.

Ecology—Vegetation

"Not even
a reed marsh
was to be
seen"

Not even a reed marsh
was to be seen.

Waste ground
your only
living place.

All lands lay...
as an ominous silence...
a desert of silent tracks.

In all the lands
there was
      no vegetation.

The great fields and meadows
    produced no grain;
the fisheries
    no fish;
and the watered gardens
    produced neither honey
    nor wine.

Your plains where grew
the heart-soothing plants
grow nothing but the
    reed of tears.

Flood-water did not flow...
Earth's womb did not give birth,
No vegetation sprouted.

My tamarisk,
that drinks not water
in the orchard bed,
the crown of which
forms not leafage
in the desert,
my poplar,
that has no joy
of its watering ditch,
my poplar pulled up by the roots,
my vine that drinks no water,
in the garden bed.

The reeds were removed.

Vegetation became too scant...
Earth clamped down her teats,
   no vegetation sprouted,
   no grain grew.

A day of darkness and gloominess...
   there hath not been ever the like...
The garden of Eden...
   a desolate wilderness.

One reed standing all alone.

Plate 1

Plate 2

Plate 3

Plate 4

Plate 5

Plate 6

Plate 7

Plate 8

Plate 9

Plate 10

Plate 11

Plate 12

Plate 13

Plate 14

Plate 15

Plate 16

Plate 17

Plate 18

Plate 19

Plate 20

Plate 21

Plate 22

Plate 23

Plate 24

Plate 25

Plate 26

Plate 27

84006

Plate 28

Plate 29

Plate 30

Plate 31

A lone reed was shaking
the head in grief.

Oh holy reed
no more!

Ecology - Animals

"The mantles
of radiance
were lost"

The mantles of radiance were lost.

The remaining offspring of living things
was tiny...
An evil wind will blow,
and the vision of people
and living things
will be obscured.

The waters and their denizens are afraid.

Therefore
shall the land mourn,
and every one that
    dwelleth therein
shall languish,
with the beasts of the field,
and with the fowls of heaven,
yes the fishes of the waters also
shall be taken away.

The rivers – their flood waters
would not bring overflow;
The fish...would lay no eggs
in the canebrake,
The birds of heaven would not
build nests in the wide earth...
Plants and herbs, the glory of
the plain, would fail to flower.

No foot of man
shall pass through the land,
nor foot of beast
shall pass through it,
neither shall it be inhabited.

Fields are desolate
without man or beast.

The freshwater fish could not spawn...
The birds of heaven not base their nests
on the broad earth,...
in the fields the tilth could not sprout
the mottled barley;
in the desert, its verdant spots
could not let grass and herbs grow
in the orchards, the broad trees...
could not bear fruit...
The wildlife...
could not lie down in their lairs
or settle on their perches,
the wild goats and asses,
the four-legged beasts,
could not be fertile.

Neither can men hear
the voice of the cattle;
both the fowl of the heavens
and the beast are fled;
they are gone.

Among the crabs
of the river, set up a lament,
    for they are gone.
Among the frogs
of the river, set up a lament,
    for they are gone.

Ecology - Agriculture

"In the
fields
the surface
had become
bad"

In the fields
the surface had
become bad,
clay clove to the skin.

The land be greatly polluted.

The dark pastureland
    was bleached,
The broad countryside
    filled with alkali.

The land has been shaken off
like a dried fig on to the ground...
Below, the water from the depths
is bolted,
    it does not flow.
That is why the dark ploughland
    has whitened,
That is why in the pastureland grass
    does not sprout.

Ecology - Laments

"My tears
ran down
my cheeks"

All life...had returned to clay.
The flood-plain was flat as a roof....
I bent down, then sat, I wept.
My tears ran down my cheeks...
Areas of land were
    emerging everywhere.

The harvest is finished...
    we are not saved.

Figure 11

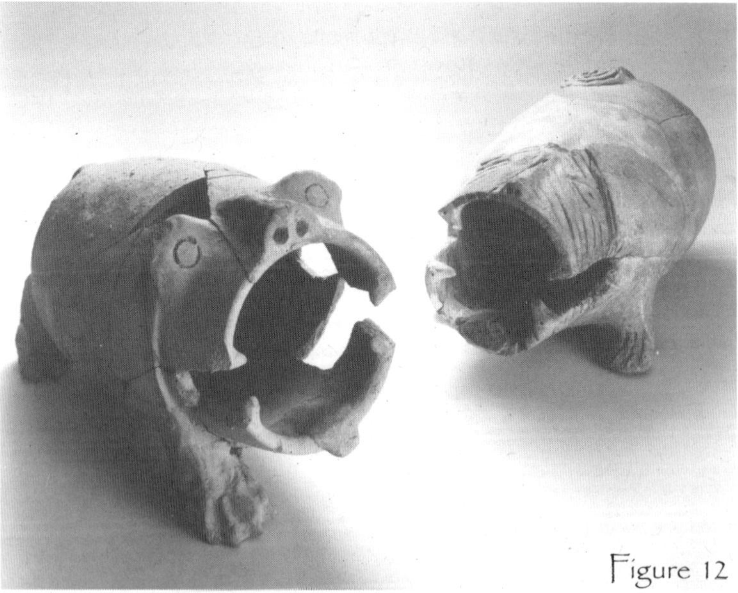

Figure 12

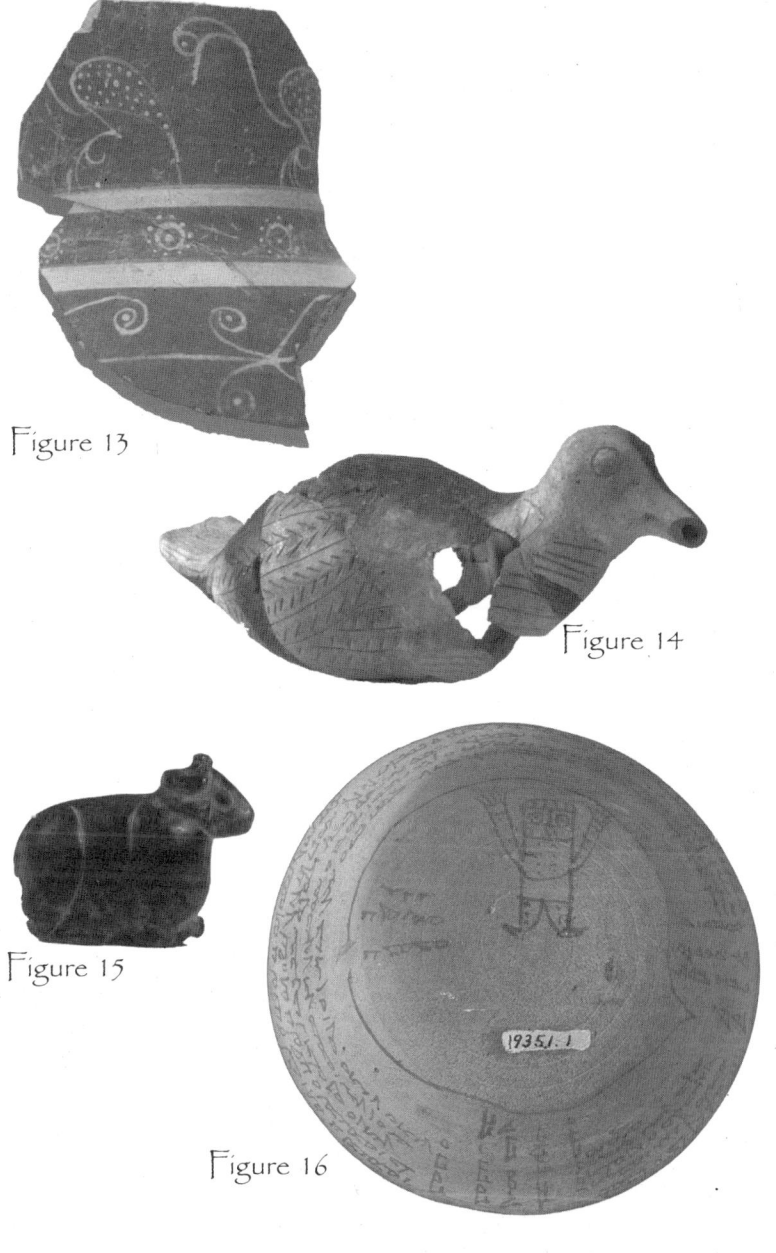

Figure 13

Figure 14

Figure 15

Figure 16

1935.1.1

Essay Two

Human Rights Issues in the Iraqi Marshlands: A Case for Genocide

Baroness Emma Nicholson of Winterbourne

In August 1988, an infamous weapons of mass destruction incident destroyed the lives of thousands of Iraqi citizens. This "massacre of the innocents" marked the first real impact of Saddam Hussein's brutality on Western consciousness. This event represented a particularly acute phase of the continuous destruction of the inherent human rights of the Iraqi people in the post-revolutionary period. There were of course countless victims tormented, betrayed and slaughtered before 1988, both inside and outside Iraq. Saddam Hussein's obsession with violence, destruction, and pan-Arab power can be gauged from his financial prioritization: in 18 of the 21 years in which he ruled Iraq, around 75% of the country's budget was spent on weapons. That means that $16 or so billion was spent year after year by one of the world's most aggressive tyrants. In consequence, the invasion of Iran which cost almost a million lives was succeeded by the invasion of Kuwait. But I suggest that the massacre of 1988—that national crime identified globally through visual media—clearly demonstrated the monstrosity of the crimes that Saddam's regime was prepared to carry out to maintain his grip on power at any cost to the Iraqi people. And at that time, in September 1988, I called immediately for the establishment of an international tribunal to try Saddam and his key officials for genocide.

A crime of similar magnitude but more drawn out and implemented during the sanctions period between 1991 and 2003 was the genocide against the people of the Iraqi marshlands. You may recall that in 1981 a pamphlet published by Saddam's foster-father stated that there were three groups whom God should not have created: the Persians, the Jews and the flies. He said the Persians were animals God created in the shape of humans, that the Jews were a mixture of the dirt and leftovers of diverse peoples, and that the flies were a trifling creation who we do not understand God's purpose in creating. His foster-son, Saddam Hussein, continued in this vein by defining the people of the marshlands as being lower than animals, as rats and as the scum of the earth. It's well know that when you define a part of humanity in such terms, you have already identified them as being as worthy of destruction and worthless in terms of care.

Genocide against the people of the Iraqi marshlands took place between 1991 and 2003. This genocide was in turn part of the constant military and police oppression against the Shiite majority of Iraq, which again was a ghastly part of the continued brutalities against almost every one of the diverse communities that had historically made up modern-day Iraq. Even some members of Saddam's own tribe did not escape unscathed; nor did his cabinet. By 1990, the whole of Iraq—as I was told by one escapee whom I met—had been turned into a modern concentration camp.

Certainly the farming communities of the Mesopotamian marshlands drank deep of the poison of Saddam's hatred expressed against them through the purposeful destruction of their identity, their culture, their livelihoods, their families, their villages, their towns, their cities, and their way of life. In essence, genocide is the intent to destroy in whole or part, a national, ethnical, racial or religious group, whether by killing, causing serious harm or by inflicting conditions calculated to bring about the physical destruction of the group. The marsh people have lived in the Mesopotamian marshlands for over five millennia and had developed a unique water-based way of life. They comprise a number of different tribes, but they all share a common culture, language, religion, and set of customs; and they can clearly be identified within the meaning of 1948 Genocide Convention as a "distinct ethnical group."

In the early 1980's, the regime of Saddam Hussein launched devastating military attacks against the Shiite Muslims of southern Iraq, killing tens of thousands of civilians. After the uprisings of 1991, the assaults were intensified; indiscriminate mass executions were carried out, killing tens of thousands more. The marsh people were targeted specifically. All of their cities, towns, villages, farms, and individual dwellings were attacked by aircraft or by artillery, and burned or demolished. Weapons of mass destruction were also employed. The survivors were forcibly displaced at gunpoint not once, but many times. Inhabitants of one such village, for example, cannot recall how many times the huge, long wheelbase tracks and armed helicopters arrived to force villagers to leave without their possessions, their animals, their livelihoods, or even their records.

Concurrently, the Iraq regime implemented a massive program of drainage and damming of the marshes in a deliberate attempt to wipe out the indigenous population. And in twenty years, as we now know, the marshes were reduced by ninety percent, causing one of the world's greatest and environmental disasters. The marsh Arabs have been deprived of their basic means of subsistence and the consequences have been catastrophic. Under such conditions, people simply cannot survive. Further, this situation was severely exacerbated by a government-imposed economic blockade of the marshes, and the deprivation of even the most basic medical care; thirty percent of the marsh people did not have the oil-for-food ration book, for example, and as such, were deprived even of food. Such actions, combined with continuous and concentrated military assaults, killed tens of thousands of people and caused massive forced displacement. From a population of over 400,000 just thirty years ago, only 85,000 survived the onslaughts inside the marshes, being forced into settlements where they were placed in indentured labor. A similar number escaped to a destitute life in cities such as in southeast Baghdad, or fled across the border in thousands to become squatters in Iranian refugee camps. All these actions have brought the marsh Arabs to the verge of extinction. Indisputably, it demonstrated genocidal action with reference to the Geneva Convention.

After one visit to the marshes in September 1992, I reported to Parliament in Britain: "Saddam has stepped up his onslaught on the marshes themselves." At that time, I traveled through marshes still smoking from ground launch bombardments; recently rebuilt villages had been destroyed and small rice plots burnt. I reached the heart of the marshes by boat, one mile from Saddam's frontline. There I found people starving—desperate people drinking filthy water and eating contaminated fish. They had fled villages under assault by Saddam's forces. Many refugees in a similar state had made the dash across the border into Iran. But to make even that crossing, these desperate individuals must have braved mined waters and a line of Saddam's soldiers. Following another trip, shortly afterwards I reported back Parliament:

"This week I visited Iran and Iraqi marshlands for the sixth or seventh time. Let me make a brief report on the situation since the implementation of the 'no-fly zone.' I have to tell you that the situation is critical. While the skies above the marshes are clear, on the ground there is revenge…Saddam Hussein has now put half his military forces into the marshes. The dangers are considerably heightened and the results can be seen even from just inside the Iraqi border. There are great black smoking areas stretching into the water where one or more missiles landed perhaps an hour earlier...If they land on a town, they demolish about 20,000 houses that go up in flames. Food stocks have been removed and taken north of the 32nd parallel. That is the first thing that Saddam did when he invaded Kuwait, he took out all the food. The farms and the marshes had been burned, the villages are no longer self-sufficient in food, but they are blockaded so that they cannot get out to the towns to try to find food on the black market. The towns are filled with soldiers. Tank divisions each containing 45 tanks are based in different towns in the marshes. Missiles and missile launches makes nonsense of the claim that it's impossible to burn towns and villages from the ground. Assault boats carrying 30 or 40 armed men (the boats in which I travel carry a maximum of four or five people) assault the towns and villages each day. The troops stationed 30 kilometers outside the marshes are coming in every day and carry out their remorseless attacks: burning, shooting, killing defenseless people and destroying whole villages and towns. I talked to a man with his hand blown off, a young farmer, only twenty years old. In the towns and villages around his area had lived 25,000 people. Five days ago, the area was emptied by 20,000 armed men. Sewage dumping has further contaminated the water, [the waste] containerized and brought in from cities around the marshes such as Basra. I saw the evidence of the drainage of the marshes. The water level has been reduced and the roots show... Medicine is not available, there are no doctors, grave shortages persist in the supply of food, water buffalo are disappearing through lack of water, the muddy, fish-stinking water and malarial air are all that is left for people to live on. If that is not genocide, the word means nothing."

In view of the enormous scale and deliberate nature of this action, combined with the documentary and videotape evidence of the planning and ordering of the assaults against the marsh Arabs, there is an incontrovertible case that the Iraq regime had genocidal intent. They had resolved to destroy the people, and this was largely achieved through the near total extinction of the marsh Arabs. Under the Genocide Convention, countries undertake to prevent or punish genocide, and to fulfill this obligation, urgent action has to be taken once genocide is soon noticed, to bring the perpetrators to justice.

I should explain that there is a difference between massacres and genocide. A massacre is when you so dislike a part of a population that you do your very best to harm them in any way. But genocide is when you try to destroy the entire population so that they cannot continue to survive. I think that the coming trials of Saddam and his key officials will demonstrate beyond any doubt that there was the intent, that there was the action, and that this was, therefore, truly a case of genocide.

It seems that the marshland society was a rare early example of an ecosystem in which humans had so fine-tuned the laws of supply and demand that they were successfully balanced in such a way that serious economic outputs (in terms of fish, dairy products, furniture and other products from reed pulp, etc.) were constantly produced and sold by every family. The marsh people are highly industrious, producing food with a wide variety of outputs, not just for themselves and their families, but for the wider world within and beyond Iraq's borders. And to do this, they successfully harnessed the Tigris and Euphrates to irrigate their grazing land and crops. And as the families have shown me over the years, men, women and children both fished and farmed with each family of around ten people, holding perhaps thirty to seventy high milk yield water buffalo, marketing the dairy produce. These water buffalo are very unusual in that they respond to the calls of their individual farmers; if you stand beside farmers and they're calling their water buffalo in the marshes, from a huge herd of water buffalo, you will see the small number coming toward the individual owner. And the families would produce two or three boatloads of fish, at least, each week. Together with wild fowl being killed and horticultural produce being harvested, as well as palm products and cane outputs, I would suggest that Adam's and Eve's idleness in the imagined Garden of Eden was unknown in the realized lower Mesopotamian marshlands, which were essentially a hive of industry. Indeed, so famous were the marshlands and the quality of their agricultural outputs historically, that the neighboring Elamites from Persia, in what is now southwest Iran, regularly invaded Mesopotamia to gain agricultural land, water and produce, whilst the Mesopotamians in turn, as regularly, invaded the kingdom of the Elamites to gain its famous mountain range mineral wealth. So the Mesopotamian marshlands had been famed for food production, food processing, and food distribution since time immemorial.

Indeed, when I first met the marsh people in 1991, their primary concern, as they were being forcibly displaced by the republican guard, was the maintenance of their herds of water buffalo. Driven at gunpoint towards a drained section of the marshlands where I found them, they talked to me about the water and grazing they so urgently required so that the cattle could survive despite the traumatic situation all were in. No idleness, but anxious and hardworking farmers, whose lands and irrigation systems were being destroyed. I sought immediately to get the world to listen, and I visited and revisited the marshlands to explain the story and tell the world this tragedy was happening. Meanwhile, finding Iraqi doctors, teachers, engineers, sanitation and water engineers, and others amongst all the displaced refugees, made it possible immediately to set up the AMAR International Charitable Foundation in September 1991. AMAR of course means "rebuild" in Arabic. Books, pamphlets, reports, photographs, films and articles in the media followed. Key publications included a book I wrote 1992, *Why Does The West Forget*, and a groundbreaking report that we commissioned from Exeter University, financed by the ordinary people who supported our work and supplemented by funding I raised from the U.K. government and World Wildlife Fund for Nature. I visited Geneva to gain support from the International Unit of Nature Conservation. And as a member of the court of Exeter University, I turned to the Vice Chancellor for assistance to put in a case for the destruction of the Iraqi marshlands to the academic world.

As a politician, I had already gained the attention of the U.K. Foreign Secretary and Prime Minister, and together we enlisted former U.S. counterparts with the support of our Ambassador there. With the concurrence of the French Foreign Minister, the organization was put in place so that since in 1993 each detail of the marsh destruction was being visually monitored and recorded from the fly-overs. The French withdrew after a short time. AMAR's scientific report was launched in the House of Commons in 1994. I addressed a specially convened meeting of the U.N. Security Council in New York around that time. My message was that the destruction was not for land reclamation—for agriculture as Saddam claimed—but was geno-cide-based and that the destruction could and should be immediately stopped, and indeed was actually reversible if actions could then be taken. Alas, the politicians could not bring themselves to act. In AMAR we fol-lowed that report with others, notably a book published on the web in draft in 2001, hardback 2002, and in paperback 2003. Full scientific data about the demise the marshes, therefore, has been on the record in front of the scien-tific community since 1992.

We brought good journalists to the area as often as possible, and in consequence, films were aired by the BBC, ITV, NBC, CNN, as well as pri-vate productions which we commissioned and financed such as *Saddam's Killing Fields*, a documentary shown by the Discovery Channel. AMAR held photographic exhibitions in a gallery in London, in the United Nations in New York, and any other location that we could find. AMAR has, therefore, built up an extensive, consistent and detailed record of the systematic onslaught by Saddam's regime on the people of the wider Mesopotamian marshland environment. The point is that from 1991 right up to and includ-ing April 2003 and subsequently, most of this information was, by one means or another, successfully put into the public domain at the time when it was happening. But as we know today, telling the story wasn't enough. Since the assaults were not curtailed and the world did not wake up in time, the plight of the people worsened daily and the slaughter continued. The towns, the villages, and the farms were burnt and destroyed. Sickness was ripe and the flood of refugees and displaced people did not abate.

From a staff of a handful of Iraqi doctors in 1991, AMAR grew rap-idly to try to meet some of the needs of those suffering people. Today, AMAR has been back working in Iraq since April 2003, at which time I and a medical team crossed the border from Iran to again set up our work. We have a staff of over three hundred people, including doctors, nurses, teachers, water engineers, and civil engineers for helping the new Iraqi ministries. For example, the health ministers for Iraq helped create a master-plan to provide primary health care to 100,000, not only in the marshes but also throughout the entire country.

AMAR's work, however, has never been solely health related. We have a careful mission statement. The object of AMAR is to recover and to sustain professional services and primary healthcare, education, and basic sanitation needs to communities living under stress, in war zones, and other areas of civil disorder and disruption.

AMAR has a number of achievements to our credit. We have provided emergency medical aid, community health clinics (at the moment, quite a number of them), health post networks, drug storage, food distribution, many thousands of tons of food and clothing, health laboratories giving full scale complete testings, preventative medicine, even health volunteers (we've trained 900 women of each of whom take on 50 families for a continuous health-training program), ambulances, and scientific surveys. We have had a scientific committee in the marshes since April 2003; we sent the first mission in, brought out the first results, and have continued to be involved in constructing water supply systems, water purification plants, surface water collection, sewage treatment plants, garbage collection, and individual house, city and school latrines. We have also built schools and kindergartens in addition to our many clinics. We have provided energy sources from small to large-scale electricity generation, and have continued to be advocates for people who have found it difficult to be heard by the outside world.

Lately, we've been pleased and delighted to facilitate and guide international visitors to the marshes, such as the USAID scientists, not once, but many times. We've carried out extensive seed distributions and have worked with FAO, IOM, and other international agencies. We now have a committee composed of scientists from several regional universities with whom we are working on a daily basis to carry out scientific surveys of the marshes on an ongoing basis. We supply internet services to the students in the universities and regularly employ them to carry out the field work.

The results of our surveys are generally published as soon as we can manage it. We carried out our first survey of the residual community of 5,000 marshland people in April to June 2003. And that was when the full extent of the horror of those who were forced to stay behind became exposed. We've conducted three more huge surveys of the inhabitants of the marshlands who survived behind the lines of the Iraqi army as it got nearer and nearer to the Iranian border, as well as a survey of those refugees returning from Iran.

Indeed, with our long history of establishing working partnerships with the marsh peoples in the time of their trauma, we have become perhaps a part of the marshland civil society as well; in other words, an esteemed friend of tribal leaders, villages and their families. AMAR works always under the authority of the local governance and national ministries, but essentially the heart of it is that we work at the request of the local people.

Marshland people have always had a strong adherence to democracy and the fundamental freedoms that they have historically enjoyed by being a valued and special part of the wider Iraqi nation. Marsh people had a historic and valuable role in food processing and distribution, rearing healthy families within a society notable for its holistic, self-sustaining way of life. I suggest that such excellence under Saddam could not really be allowed to survive. Had the marshes really been either the Garden of Eden, that ideal of idleness, or a distant, poor and sparsely populated sub-region, surely the marshlands and her people would never have attracted Saddam's attention. But this was an independent, successful, hardworking farming community who stood up for their rights, fought for their rights, and for their place in Iraq.

The case, therefore, is genocide—that crime of crimes against humanity. The perpetrators will be held accountable before the Special Court. Already, meanwhile, thanks to actions of many, much of the marshes have already been partially re-flooded with large volumes of water being replaced. And I've had the joy of watching that water come back from a viewpoint on the ground and the air. But there's much more to be done than just supplying water to get this historic community back to work; health and education as well as physical land restoration are all required.

Access to health, to education, the right to life and property are of course fundamental human rights as laid out in the United Nations Charter of the 10th of December 1948. So how does that charter compare with today's plight of the marsh people? Well, 97% of the 85,000 people we surveyed in May 2003 had had no access to public health for over three decades, and 87% are illiterate or non-numeric because of the withdrawal of education facilities. And even now today, many months after the toppling of Saddam, a hundred percent of those people are suffering from sickness (diarrhea and from other preventable and easily treatable illnesses). And all these people were forcibly displaced, every single one of them against their fundamental rights, in some cases, up to seventeen times (!), loosing their houses, land, water, livestock and livelihood, again and again, always without their intrinsic right to safe drinking water and safe waste disposal.

Here is the *Universal Declaration of Human Rights* to which I am referring with articles pulled out: "Everyone has the right to life, to liberty and security of person." "Everyone has the right to own property alone as well as of association with others and no one shall arbitrarily be deprived of his property." Property rights do not exist just because you happen to have a piece of paper in the modern world. Property rights exist because you live there. "Everyone has the right to work, to free choice of employment in just and favorable conditions of work. The right to take part in the government with equal access to public services." And we all have rights of standards of living adequate for our health and well-being and necessary social services such as food, clothing, housing and medical care. Security in times of sickness disability, whether in old age or other lack of livelihood, as well as a right to education are all fundamental rights that have been destroyed for many of the Iraqi people. As the latest Arab Human Development Report demonstrates clearly, the people of the Arabian Peninsula and North Africa believe strongly in justice, democracy and in the rule of law. The Iraqi people believe that the will of the people should be expressed to the establishment of democratic institutions such as Parliament and elected ministers. The occupying powers came in to free the Iraqi people from terribly tyranny. And we, with our own deep commitment, to helping the Iraqi people to establish democracy, are committed to that goal.

Whether or not you believe that war was the necessary last resort instrument to bring about the deserved destruction of the tyrannical regime —I did and do—we now all carry a burden of responsibility towards the Iraqi people that history will not forget. It is a burden that should force us not to compete to help the marsh dwellers or other people of Iraq, but instead to listen to them and to facilitate with all the energies and will and talents and skills we have, the ongoing restoration and development of stable and effective democratic institutions and the rule of the law. And it is in this context that we must think about the work in hand, that the so excellent Harvard conference has put in front of us. The task is bringing back help, education, the right to life, and land restoration to the marsh people. Just as it is our duty to support the people elsewhere in Iraq with those same benefits, I as an E.U. Parliamentarian and all of us in AMAR are proud to assist the process.

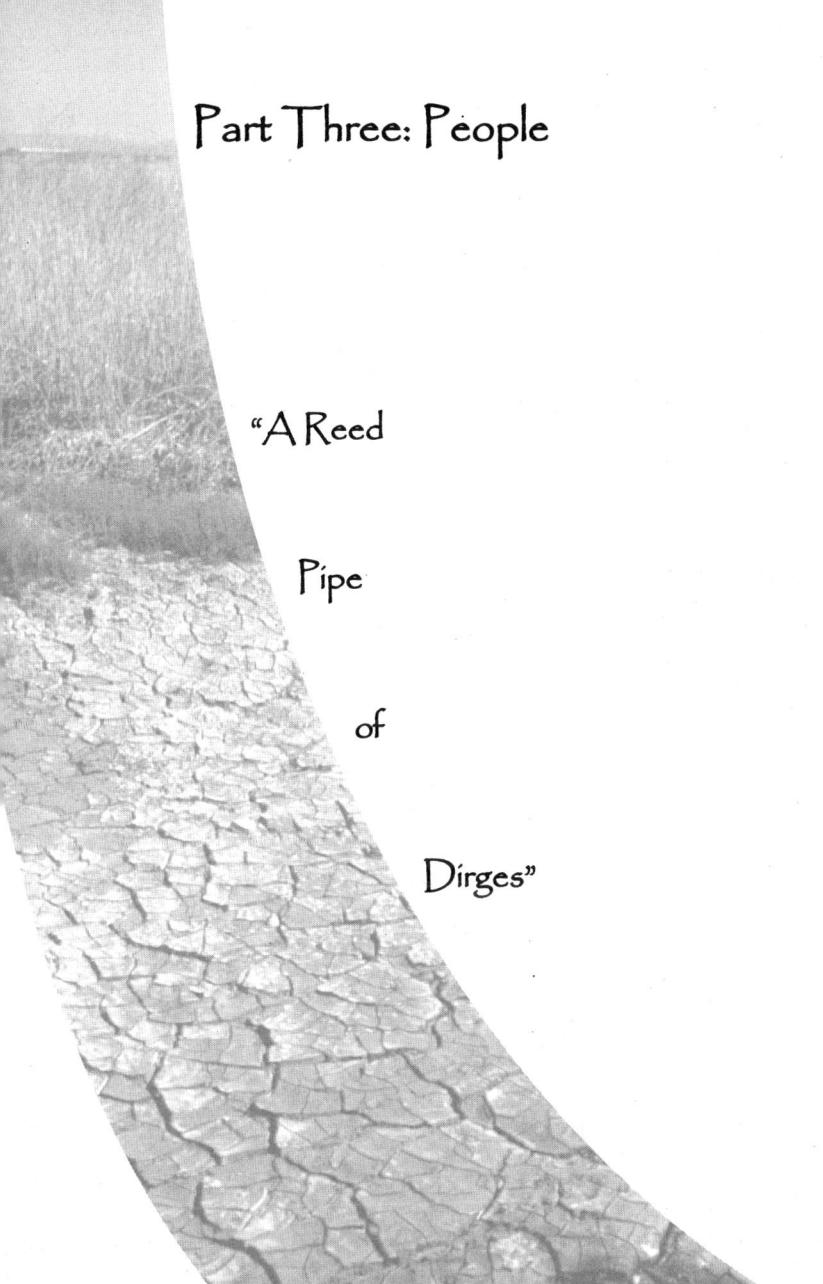

Part Three: People

"A Reed

Pipe

of

Dirges"

People ~ Saddam

"He plotted evil,
to devastate the land,
to destroy the people"

He plotted evil,
to devastate the land,
to destroy the people.

An evildoer will seize you.

The lord spewed bitter venom
at the rebel country.

The cities with fire
I burned.

Murderer, unspairing,
    ravager...
Sharp were the teeth,
    laden with poison.

For your hands are defiled
    with blood.

I will pour out my fury
    upon them in the wilderness,
    to consume them.
I lifted my hand unto them
    also in the wilderness, that
I would scatter them...
    and disperse them
    through the countries.

I sent my warriors
into the midst of the
swamps...
and they searched for
five days.

I will also water
    with thy blood
    the land.....
I have caused my terror
    in the land of the living.

Like a lone reed, like a lone reed,
   the mighty one is shaking me,
the mighty one the lord of all lands,
he of the unfathomable heart,
of the effective words,
whose commands
are not to be countermanded...
like embedded rushes,
like embedded halta-grass,
like a lone poplar planted on the riverbank,
like a dogwood tree planted on dry land,
like a lone tamarisk,
planted where there are tempests,
like a lone reed
   the mighty one is shaking me!

I killed the inhabitants...
who had been disobedient.

You have made their blood
    flow like water in the
    drains of pubic squares.
You have opened their veins
    and let the river
    carry off their blood.

Of those who had...
entered and occupied a place
    of refuge,
not one escaped,
not a rebel escaped
    my hands;
in their places of refuge
    my hands
captured them.

Hear, O earth and behold,
I will bring evil upon this people...
because they have not hearkened
unto my words, nor to my law,
but rejected it.

I shall cut off the noise
of mankind
and deprive him of joy.
Like the fire where
there once was

      peace.

That he should kill.
that he should destroy.
In the morning that he
should rain down
the extermination.

I will scatter them...and
I will send a sword after them, till
I have consumed them.

Not one among them did
    I leave,
their youths and maidens
as a holocaust
    I burned.

With corpses of their warriors
on the wide plain I filled.

People - War

"A day
of
doom"

A day of doom...
bathed the sky in blood...
scattered its people,
and till today black cinders
are in the fields,
and ever heaven's base
becomes to the observer
like red wool—thus verily it is.

Behold, a people cometh
from the north country...
they shall lay hold on
    bow and spear;
they are cruel and have no
    mercy;
their voice roareth like the sea;
and they ride upon horses,
set army as
    men for war.

I look about me:
    evil upon evil...
Whence came
    the evil things
        everywhere?

Be it known
that your city
will be
completely
destroyed.

With arrows and bows
shall men come thither.

No man should have
lived through the destruction!...
Punish the sinner for his sin,
the criminal for his crime.

Heaven cried out,
earth groaned.
Day grew silent,
darkness emerged.
Lightning flashed,
fire broke out.
Flames crackled,
death rained down.

A sound of battle is in the land,
and of great destruction.

Against the king,
    the small were hurled.

Death
dropped
    over them
    like a fog.

The land is full of bloody crimes.

The land has become dark,
    It is full of shadows.

Clouds of death rained down,
and arrow flashed lightning,
    whizzed,
the battle force roared.

They shall burn thine houses
    with fire.

People - Death

"The dead
outnumber
the living"

The dead outnumber the living!

Bitter is the wail
for her ravaged city!...
your people
    are perished.

Their myriad corpses
will reach the
base of heaven.

In thy skirts
is found
the blood
of the
souls
of the
poor innocents.

Thou shalt be fuel for the fire;
thy blood shall be
in the midst of the land;
thou shalt be
no more remembered.

The people mourn...
in all the lanes and alleys
corpses were piled,
and in all the open spaces
where the country's dances
once were held,
people were stacked in heaps.
The country's blood filled all holes,
like copper or tin in molds,
their bodies
like sheep fat left in the sun
dissolved of themselves.

Thou shalt die the deaths
of them that are slain
in the midst of the waters.

Man perishes,
   heavy is the heart.
I peered over the wall
Saw the dead bodies floating
   in the river's waters.

This river of spent souls.

And the carcasses of the people
shall be meat for the fouls of heaven,
and for the beasts of the earth.

In a desolate place, where once
he was alive; now he lies
like a young bull felled to the ground...
His mother who has lost him
to death's kingdom.
O the agony she bears,
Shuddering in the wilderness.
She is the mother
suffering so much.

People ~ Capture

"Their hands
have caught
you"

Woe lad
on the road they searched for him,
in the desert they searched for him,
they scanned the landscape,
they saw him,
they cried out, they seized him,
his girdle they untied...
the lad's thighs, they bared...
he was blindfolded and, bound as he was,
he was driven along, allowed no sleep.
Alas, lad!
Their hands have caught you.

The murderer
    with his weapons
        pursues him.

The man in front of him
threw things at him,
the man in back of him
was...clapping manacles
on his hands,
was pinioning his arms
in a shackle.

The fisherman who used to
bring you fish is
    overtaken by misfortune;
The bird hunters who used to
bring you birds were
    carried off.

On my city's fields is
       no grain,
their farmer was led off
      captive.

My bones are smashed...
My hands have been cast
    into fetters...
The lash striking me is filled
    with terror...
All day a pursuer pursues me.
At night he does not let me draw
    my breadth for a moment.

People – Refugees

"The people
scattered"

Woe!
My city has been
    destroyed—
Woe!
My house has been
    ravaged!...
The people
    scattered...

The dark-headed people
were driven off...
    into slave quarters.

The whole city shall
flee for the noise
of the horsemen
and bowmen;
they shall go
into the thickets.

(Into the land
whence none return,
the place of
      gloom.

I have forgotten mine house,
I have left mine heritage.

And the word shall come...
and great pain...
and they shall take away
the multitude.

So I went, and hid...
by the Euphrates.

And they escaped into the land.

Death pursued me...
My people weep,
I myself no longer exist.

Weep sore for him
that goeth away:
for he shall return no more,
nor see his native country.

I depart, never to return,
into the land of
darkness and gloom.

The wandering dead,
the fleeting ghosts.

People - Laments

"Laughter is
become
lament"

Laughter is become lament.

He sat facing the river,
   he wept,
The man wept facing the river.

Although
I am constantly looking for help,
    no one takes me by the hand.
When I weep they do
    not come to my side.
I utter laments, but
    no one hears me;
I am troubled;
I am overwhelmed.

Mine eyes are a fountain of tears,
that I might weep day and night
for the slain...of my people.

They bring pain,
    heartbreak for mankind.
The barges on the river bring pain,
    heartbreak for mankind.

Will I take up
weeping and wailing
and for the inhabitants of the wilderness
a lamentation,
because they are burned up.

Pity! For my sickened heart
      is full of tears and suffering...
I am beaten down, and so
I weep bitterly.
Death and trouble are
bringing me to an end.

Let screams
draw near unto heaven,
draw near unto
the netherworld;
the sum of screams
covered like a cloth
the base of heaven,
veiled it
like a linen sheet.

The men shall cry,
and all the inhabitants
of the land shall hear.

His heart was full,
he went into the barren land
    and wept;
the shepherd's heart was full,
out in the barren land
    he wept.

How long shall my body lament,
    full of trouble and disorders?
How long shall my heart
be affected,
    full of sorrow and sighing?

In the desert...
she holds not back
the flood of tears
for her husband.

The people mourn...
The old men and women
who could not leave the house
were consigned to the flames...
The country's sense vanished
the people mourn.

Your songs have been turned
into laments before you.

Grief has entered
my innermost being,
I am afraid of Death...
Death's picture
cannot be drawn.

A reed pipe of dirges—
    my heart wants to play
a reed pipe of dirges
    in the desert!

Weep, weep, wilderness weep...
My heart is piping
in the wilderness
an instrument of grief.

Tears, lament, anguish,
and depression are
lodged within me,
Suffering overwhelms me like
one chosen for nothing but tears.
Evil...holds me in its hand,
carries off my breath of life.

Figure 17

Figure 18

Figure 19

Figure 20

Figure 21

Essay Three

Experiences and Hopes of the People of the Al-Ahwar Marshes

Rasheed Bander Al-Khayoun

Issues concerning the revitalization and redevelopment of the marshes, or what we call the *al-Ahwar* of southern Iraq—the place of my childhood and youth—are of critical importance. I will concentrate my discussion in this essay on the al-Ahwar region only, recognizing of course that many other parts of Iraq were also targeted severely by Saddam's regime.

Throughout Iraq's history, many plans have been laid out to drain the marshes, sometimes with the aim of increasing tax revenues and sometimes with the aim to opening up the area as it had long been used as a refuge by political opponents. The attempts of Iraq's former regime to drain al-Ahwar had various motives. In addition to imagined political and security concerns, there were also sectarian motives since the area is almost entirely Shi'a, except for the Mandean Sabeans (a very ancient religion that appeared before even Judaism and Islam and having roots extending back to the Sumerian civilization). Drying up the waters of the marshes would also mean drying up the rural culture and heritage which could not survive outside of that watery environment, for Shi'ism has a long history in the al-Ahwar and is inscribed within the culture and literary heritage of the area. In addition to this, the state was concerned with the question of agriculture and the presence of oil wealth beneath the vast spread of water.

From my own personal experience, the people of al-Ahwar realize that they inhabit an ancient place, but they refer to hills within the marshes by the name of *ishan* without knowing that it is a Sumerian word meaning "hill," and without realizing that the marshes were once part of the Sumerian land. What I heard from my mother, which was handed down from her forebears, was that in olden times there were palaces and cities which raised God's anger and which were therefore overturned because of their lack of respect for His grace. Legend had it that long-ago there was a woman baking bread with her child beside her, and that she started using the loaf of bread to clean up her child's excrement, and from that day on there were no more palaces or cities or civilization in the marshes. In fact this story remains important in terms of its educational and religious value for it aims to emphasize the fear of God and respect for His beneficence. Interestingly, even today, the only foods which the people of the marshes consider as holy are rice and bread (for you can walk on dates or meat or apples but you should fear the wrath of God if you accidently tread upon rice or bread).

How well I remember, growing up in al-Ahwar, looking upon foreign tourists with surprise and with a degree of amusement, for their questions concerning what it meant to be living on water, or catching fish from a boat made of reeds, or what it was like to live alongside snakes crammed into a single house, or what was it like to watch the moon plunge into the broad surface of the water and other magnificent scenes to which we paid no account for they were part of our daily lives to which we were accustomed to and thus did not see as new or surprising. At that time, we considered the astonishment of strangers to be courtesy or flattery, and a pleasant response to our generous hospitality.

Is it not surprising and beautiful at the same time that a person cannot remember when he or she first learned how to swim? For after I moved to the city to study and to work I was surprised by the presence of closed-in swimming pools for teaching this skill, because I had learnt to swim at the same time as I learnt to crawl, both well before walking. Perhaps the words of the British Orientalist Lady Drower give best expression to my surprise at the existence of formalized classes teaching people how to swim in dry cities, when she wrote "the people of the marshes have many of the features of water birds, they are cheerful, enjoy jokes, and are fond of laughter, and are passionate about singing; their homes are dry even with water all around; no sooner does a child start to walk than you find him swimming in the water, they are like the birds of the water, and they will remain like that, probably to the end of time." How sad we find her optimistic phrase "to the end of time," today?

The intellectuals from among the people of al-Ahwar—from among those who continued going back to the region and who still have family remaining there—realize that global tourism is the future which beckons, and they do not hesitate to call the region the "Venice of Iraq," despite the fact that the majority of them have only seen that wondrous Italian city in photographs or on television. But this demands that certain conditions be fulfilled, the first of which is organizing the channels of water, and building hotels, both on the water and on land, and restaurants, and providing other services, etc. Two further conditions needed are that these facilities should be built with materials from the environment itself like reeds and rushes, and that tourism would be active in winter, not in summer, so that tourists could fully enjoy al-Ahwar in all its watery wonder.

The people of the region are respective to outsiders and visitors. I remember that the Iraqi state at the time of the monarchy forced opposition politicians who were Jewish or Christian to leave Baghdad and Mosul and take up residence in our region, but the state went back on this after only a few months and moved them to different places because it felt that the people of the marshland region were sympathetic towards such refugees. I affirm once more that the prospect of winter tourism would be successful after the provision of all that infrastructure that is required for development.

Tourism alone, however, is not the complete future of the region. Rearing fish and birds as well as growing plants required by various industries could be important means to ensure the region's long-term economic viability. And possibly building model villages for the people of al-Ahwar and spreading the services of the city to them would provide an attractive alternative for many of those living in big cities. Development of such model villages requires a kind of mixture between offering necessities of modern-life while at the same time preserving the environment of the marshes. All people of the marshes agree on their requirement for suitable and comfortable living conditions. As my mother told me when she still kept two cows—in spite of the draining of the region—the thirst of a cow and the duck's search for a pool to swim in reveals the falsehood in all the claims of the previous regime concerning the imagined economic benefits entailed through draining the al-Ahwar.

The people of al-Ahwar need water not just because it is necessary to raise cows or ducks or to grow rice, but rather because their spiritual need for it surpasses such pragmatic concerns by far; for draining the marshes meant putting boats out of service and preventing people from going to the beds of reeds and rushes, and thus an end to the reciting of the poetry specific to al-Ahwar and to the singing which can only be performed in that theatre of water and reeds and rushes. Draining the marshes meant the death of a way of life that people had practiced for tens of centuries, a life offering passion during work among those reed beds. It is impossible to underestimate the magnitude of the loss to the people of al-Ahwar of their water, their reeds and rushes, their fish and birds, their animals, and their feelings about their environment.

The area is now being prepared to receive an increase in the waters from the Euphrates and Tigris rivers. With reflooding parts of the central region of al-Ahwar, the floating islands of land will rise up just as the water level rises, thereby lifting up the people and the water buffalos and the guard dogs. At the edges of al-Ahwar there are once again large boats and platforms made from reeds. But for the people of al-Ahwar the solution is to develop a mechanism for regulating the floods which usually come in spring or autumn. And though it may be that such severe floods were disastrous in the past, the harm done by them can, of course, never match the destruction of draining the marshes dry.

The people of al-Ahwar have expressed to me their sense of approaching joy as they anticipate the return of water and fish and birds. Many people are aware of the possibility that the al-Ahwar may become a nature reserve under the protection of the United Nations. They see the nature reserve as a way of bringing its people within the bounds of civilization and civility, and of safeguarding the region from the bad moods of local authorities.

In the end, I think people of the marshes want their marshes back but in a better way; they want means of modern life to be made available to them; they don't want floods that disrupt their way of living; they are keen to renew their life but not at the expense of their natural environment; they would welcome any activity of tourism; and they wish for some land to be reformed so they can continue to cultivate rice. The people of al-Ahwar dream of a return to marshes filled with fishes, birds, diary cows and buffalos, while moving about on modern passage ways and living on their modernized islands, since such a dream will be in harmony with their spiritual heritages of their songs, poems and tales.

The people of al-Ahwar have simple concerns centered on their daily lives, the security of their homes, and the well-being of their families, and the continuity of the environment around them. They are pleased with any service offered to them and ready to accept any plan for the future, respectful to their interests.

Last year I received an email message in which my nephew, whom I had last seen as a child, wrote to me: "Uncle, I am Rabi', writing to you from al-Jabayish by means of the internet." The letter gave me the title for an article I later wrote in the *al-Sharq al-Awsat* newspaper: "Bringing back al-Ahwar... verdant green with internet and satellites." The water has begun to arrive and with it the swimming fish and the birds, but this return is coming with all the trappings of a technically advanced civilization.

Not too long ago, the people of the marshes feared their land was about to disappear forever. Today, however, they can see that there is a great opportunity that their previous lives and the survival of their watery land is about to be restored in front of their very eyes...

Postscript for Tomorrow

The End is Nay:
Eden Restored

The land shall be desolate;
yet will I not make a full end.

—Jeremiah

My father, may it be restored!
When? May it be restored!

—The Destroyed House

Ecological restoration is a positive process of offering hope for a better future while at the same time acknowledging an often shameful past. By addressing and correcting the sins of history, restoration becomes an act of reciprocity, important not only for improving the quality of the outside environment of nature, but also that of the internal environment of the psyche, or human nature. Therefore, more than simply being a collection of fixed end-products, restoration is a healing process of both ecological spaces and consciousness, each in obvious need of repair in Iraq. And, given that some practitioners have likened ecological restoration to a form of therapeutic gardening, what better place to engage in such than in the original garden—Eden?

The promise and possibilities of restoring the Iraqi marshes have captured worldwide attention. Searching for some positive news to report from "post"-war Iraq, the Press in particular has embraced the concept of restoring the imagined Biblical Garden of Eden, and through their coverage, has fueled the imagination and hopes of many environmentalists and ex-patriot Iraqis around the world.

Marsh restoration efforts have come to be regarded as perhaps the least controversial and in many respects most helpful of all the rebuilding projects taking place in war-ravaged Iraq. Millions of dollars directed to these efforts have already been promised (and some spent) by the donor countries of Italy, Canada, Japan and the United States. And many detailed technical plans already exist, guided by some of the world's leading wetland scientists and hydrologists, outlining how to proceed with the complicated task of restoring the marshes.

The restoration of the southern Iraqi marshes is by no means only a top-down, international effort in either war-guilt appeasement or anti-Saddam rhetoric. There is a very strong desire from many one-time marsh dwellers to bring back as much of the former environment as possible. For no sooner had the dust settled from the passage of American tanks on their route from Basra to Baghdad, than the locals had immediately begun to punch holes in Saddam's dikes, allowing the imprisoned water to flow out once more across what had been their former marshland home. This hunger for a return of, and in some cases, to, the marshes seems encoded in the remnants of the indigenous culture forced into nontraditional dryland farming for over a decade.

The NGO Iraq Foundation's "Eden Again Project" has been instrumental in brokering international knowledge and marrying it to local wisdom to bring about the first marsh restoration efforts on the ground. Benefiting from an atypically wet winter and consequently large spring freshet, what had perhaps originally been planned as a small demonstration project ended up far exceeding expectations in terms of both the size of the area reflooded and revegetated and the amount of waterfowl successfully recolonized. These carefully planned scientific efforts combined with the continued opportunistic release of water by locals has meant that nearly 20 percent of the former areal extent of the marshes has returned.

Hydrological modeling has suggested that given the current amount of water released by upstream stakeholders (particularly Turkey), it may eventually be possible to reflood up to 40 percent of the original marshes. This bodes well for restoring the wildlife of the marshes, the working precept behind wetland restoration very much being "build it and they will come." Whether this is enough, however, to restore the sociology of the marshes is much more difficult to answer. Such topics were explored at the Harvard conference I organized in 2004 titled "Mesopotamian Marshes and Modern Development: Practical Approaches for Sustaining Ecological and Cultural Landscapes." The bottom line is that it is impractical, unrealistic and condescending to expect the returning Ma'dan to assume a museum lifestyle to satisfy our own romanticism.

The challenge in the comprehensive restoration of the Iraqi marshes then becomes how best to interface a desire by some marsh Arabs to return to a semblance of their former lifestyle and at the same time keep the benefits of modernity to which they are entitled (such as health care, communication, sanitation engineering, education, transportation, industry, agriculture, and tourism to name but a few). In the end, the restoration of the marshes and their inhabitants will be judged a success only by approaching that restoration within a framework of sustainable development.

And though those involved in the marshes hold variable opinions as how best to achieve such a sustainable restoration, all share an unbridled hope and strong belief that the once "desolate" marshes are far from being at a "full end," and that this onetime inhabited paradise can certainly "be restored" as the ancient Biblical quotations and Mesopotamian inscriptions portend. For though the "destroyed house" of the marshes has been severely damaged for its residents, it can and indeed *is* being rebuilt through the cumulative efforts of many restorationists operating in an optimistic spirit captured by this concluding prophecy from Isaiah:

> They...shall build the old waste places...shall raise up the foundations of many generations; and shalt...be called the repairer of the breach, the restorer of places to dwell in.

And the living creatures ran and returned
as the appearance of a flash of lightning...
This land that was desolate
is become like the garden of Eden...
Plant that was desolate...
so shall the waste be filled again with flocks.

—Ezekial

The land is the Garden of Eden before them, and behind them a desolate wilderness

—Joel

The land is restored.

—Enki and the World Order

Appendix

Harvard University's Collections of Near Eastern Artifacts and Archive of Early Marsh Arab Photographs

The Semitic Museum is Harvard University's museum of Near Eastern archeology. The centerpiece of its Mesopotamian collection comes from excavations sponsored in association with the Harvard's Fogg Museum between 1927 and 1931 at the site of Yorghan Tepe in northern Iraq. Occupied from around 5000 BCE to about 400 CE, the settlement at Yorghan Tepe was called Nuzi during the middle of the second millennium BCE, when it was a provincial town in the small Hurrian kingdom of Arrapha, whose capital is located today under the center of nearby Kirkuk. Hurrians occupied the northern periphery of the Sumerian-Babylonian heartland for many centuries. Already at the end of the third millennium BCE some officials in Sumer—which included the southern marshlands—had Hurrian names. Hurrians adopted many aspects of their Mesopotamian neighbors such as writing in Babylonian cuneiform, as the nearly 5,000 tablets found at Nuzi demonstrate.

Other material housed in the Semitic Museum of Mesopotamian or contemporaneous origin includes several thousand cuneiform tablets in addition to those from Nuzi, as well as cylinder seals, inscriptions, coins, pottery, glass and figurines.

Among the staggering half-a-million photographic images archived in the Peabody Museum of Archeology and Ethnology at Harvard University are 120 small black-and-white snapshots. These images were taken during the Field Museum of Chicago's 1934 expedition to southern Iraq and were donated to Harvard University by the expedition leader, Henry Field, one of the most prominent anthropologists of his day. The purpose of the expedition was to conduct an anthropometric survey of the inhabitants, in particular members of the Al-Bu Mohammed coalition of tribes in the southeastern part of the marshlands. The bulk of the photographs thus resemble criminal mug shots of mostly men in profile, and the published reports contain dozens of pages of dense tables filled with copious measurements of faces, noses and skulls, all directed to a then believed-in concept of what we today would refer to as racial stereotyping.

During the expedition, a number of photographs were also taken of what were referred to as "life scenes." These images were of reed houses, canoe building, craft making, fishing and hunting, as well as landscapes. As such, the photographs in the present book—many of them taken by principal expedition photographer Richard Martin and published here for the first time—together with recently published images from the 1890s and others from well known adventures who visited the marshes in the 1950s and 1970s, provide a rare glimpse of a timeless world that has now all but vanished. What is most amazing is the truly timeless nature of this unique cultural landscape; in other words, if one collected all the images spanning nearly a century and shuffled them all together it would be difficult if not impossible to distinguish when each photograph had been taken. And what is most depressing is that so much of it is now a lost landscape, never—even if the most optimistic restoration plans come to fruition—to be seen again on the same scale and in the same way.

Inscription and Quotation Key

Inscription Sources

Brooks, B.A. 1921. *A contribution to the study of the moral practices of certain social groups in ancient Mesopotamia.* W. Druglin.

Chjiera, E. 1938. *They wrote on clay: The Babylonian tablets speak today.* Univ. Chicago Press.

Cooper, J.S. 1986. *Sumerian and Akkadian royal inscriptions. Vol. I. PreSargonic inscriptions.* Amer. Oriental Soc.

Dalley, S. 1989. *Myths from Mesopotamia: Creation, the Flood, Gilgamesh, and others.* Oxford Univ. Press.

Harper, R.F. 1904. *Assyrian and Babylonian literature: Selected translations.* D. Appleton and Comp.

Heidel, A. 1949. *The Gilgamesh epic and Old Testament parallels.* Univ. Chicago Press.

Heidel, A. 1951. *The Babylonian Genesis.* Univ. Chicago Press.

Hess, R.S. and D.T. Tsumura. 1994. *"I studied inscription from before the Flood": Ancient Near Eastern, literary, and linguistic approaches to Genesis 1-11.* Eisenbrauns.

Jacobsen, T. 1987. *"The harp that once…": Sumerian poetry in translation.* Yale Univ. Press.

Kramer, S.N. 1963. *The Sumerians: Their history, culture, and character.* Univ. Chicago Press.

Leick, G. 2001. *Mesopotamia: The invention of the city.* Penguin.

McCall, H. 2001. *Mesopotamian myths.* British Museum Press.

Mendelsohn, I. 1955. *Religions of the ancient Near East: Summero-Akkadian religious texts and Ugaritic epics.* The Liberal Arts Press.

Mercer, S.A.B. 1919. *Religious and moral ideas in Babylonia and Assyria.* Morehouse.

Roberston, H.S. 1900. *View of the past from Assyria and Babylonia.* George Bell and Sons.

Sanders, N.K. 1971. *Poems of heaven and hell from ancient Mesopotamia.* Penguin.

Scharder, E. 1888. *Inscriptions and the Old Testament.* William and Norgate.

Young, G. (photographs by N. Wheeler). 1977. *Return to the Marshes: Life with the marsh Arabs of Iraq.* Collins Books.

Photographic Key

Museum Replicas:

The front cover is a composite of three images: the luxuriant wetlands at the top are part of the restored Hula swamps in Galilee that—interestingly and not without a bit of irony—were drained by the Israelis in the 1950s and which at one time had had their own indigenous population of marsh Arabs; the desiccated wetlands along the side are from the Las Vegas Wash, another example of wanton destruction and fledgling restoration of rare desert wetlands; and the warrior is a reproduction from a wall frieze from Ninevah in the British Museum of, appropriately, another despot from northern Mesopotamia—in this case Assurbanipal, last of the kings of Assyria (650 BCE)—wrecking havoc yet again on dwellers of the south.

The images on the back cover are of reproductions of two of the most widely distributed of all Mesopotamian statues: the one in white is a scribe from the Louvre Museum (Susa, circa 2350), and the black one is of Gudea (Prince of Lagash, circa 2050 BCE) from the University of Pennsylvania Museum.

The images of hand-held cuneiform tablets before each of my four essays are also of reproductions: the first for the Preface is from the city of Ugarit along the Syrian coast (no information provided, but probably circa 1300 BCE; obtained at the site); the second for the Prologue is from the Sumerian city of Shuruppack, circa 2600 BCE (Louvre Museum); the third for the Introduction is from the Babylonian city of Nippur, circa 650 BCE (University of Pennsylvania Museum); and the fourth for the Postscript (appropriately only half-scripted since the restoration work is also incomplete) is from the Sumerian city of Ur, circa 2000 BCE (IBSS mail order).

The small images at the end of quotation groupings are also from replicas, three described above (white scribe statue, King Assurbanipal frieze, and Ugarit cuneiform tablet), in addition to a bull that was once part of a lyre found in the Royal cemetery at the important Sumerian city of Ur, circa 2500 BCE (University of Pennsylvania).

Marsh Dwellers:

These photographs were all obtained from the Peabody Museum of Archeology and Ethnology at Harvard University and, as described in the Appendix, were taken during the Field Museum of Chicago's 1934 expedition to southern Iraq, and donated to Harvard University by the expedition leader, Henry Field (see Al-Dewachi 2004).

I have arranged the photographs in a general sequence progressing from panorama to intimate detail, beginning with the marsh landscape (Plate 1), boat building activities (Plates 2-4), heading out on a boat journey (Plates 5-8), moving deep into the marshes (Plate 9) and along a river orchard (Figure 10), approaching (Plates 11-16) and then entering (Plates 17 & 18) a village, a tribal meeting (Plates 19 & 20), Ma'dan fishing with a spear (Plate 21) and an expedition member boar hunting (Plate 22), village women cooking (Plates 23-25), goat milking (Plate 26), and displaying crafts (Plate 27), meeting of men at edge of the marsh (Plate 28), children playing in a baking tray (Plate 29), women and girls shown in fancy dress (Plate 30), and a turtle fighting match (Plate 31).

Mesopotamian and Contemporaneous Artifacts:

(c/o James Armstrong, assistant curator of collections at Harvard's Semitic Museum):

Figure 1. Alabaster tablet bearing an inscription of the Assyrian king Adad-nirari II (ca. 900 BCE).

Figure 2. Clay tablet from Puzrish-Dagan recording the payment of provincial taxes (21st cent. BCE).

Figure 3. Clay tablet from Uruk (2nd cent. BCE).

Figure 4. Unopened sealed clay envelope from Girsu that contains a tablet recording the disbursement of grain rations (21st cent. BCE).

Figure 5. Clay tablet from Girsu recording the receipt of commodities, still partially encased in its envelope (21st cent. BCE).

Figure 6. Very early clay tablet from Uruk. The impressions are numerals; to their right are inscribed a plow and two hands, probably representing a personal name (ca. 3000 BCE).

Figure 7. Front board of a terracotta model chariot from Larsa depicting a king (left) standing before a seated deity (early 2nd mill. BCE).

Figure 8. Terracotta plaque showing Ishtar, Mesopotamian goddess of love and war, riding on a lion (early 2nd mill. BCE).

Figure 9. Terracotta head from Larsa depicting the Babylonian demon Pazuzu (early 2nd mill. BCE).

Figure 10. Terracotta offering stand from Nuzi. The base of the stand is in the shape of a house with windows and parapets. A separate bowl would have rested on top (14th cent. BCE).

Figures 11 & 12. Ceramic votive vessels from Nuzi, crudely fashioned to represent lions (14th cent. BCE).

Figure 13. Fragment of a painted ceramic cup from Nuzi showing waterfowl (14th cent. BCE).

Figure 14. Ceramic votive vessel from Nuzi in the shape of a bird (early 2nd mill. BCE).

Figure 15. Stone amulet from southern Iraq in the shape of a cow (late 4th mill. BCE).

Figure 16. Ceramic bowl from southern Iraq bearing a Mandaic incantation and the image of a demon with upraised arms in ink (mid-1st mill. CE).

Figure 17. Fragment of a stone bowl bearing an inscription of the Babylonian king Nabonidus (6th cent. BCE).

Figure 18. Stone weights from Nuzi in the shape of trussed ducks (14th cent. BCE).

Figure 19. Terracotta female figurine from southern Iraq (late 3rd mill. BCE).

Figure 20: Terracotta figurine from Uruk depicting a woman nursing an infant (mid-1st mill. BCE).

Figure 21. Alabaster statuette from Girsu depicting Ur-Nanshe, king of Lagash (25th cent. BCE).

Bibliography

*Al-Dewachi, O. 2004. Field photography: The marsh Arabs of Iraq, 1934. Exhibition brochure from the Peabody Museum of Harvard University.

Anon. 1995. *Mesopotamia: The mighty kings.* Time Life Books.

Anon. 2003. Building a scientific basis for restoration of the Mesopotamian marshlands. Iraq Foundation Eden Again Project.

Anon. 2005. The new Eden project: Final report. Italian Ministry for the Environment and Territory and the Free Iraq Foundation.

Anon. 2005. Iraq marshlands restoration program action plan. U.S. Agency Internat. Develop.

Balakian, P. 2003. *The burning Tigris: The Armenian genocide and America's response.* HarperCollins.

*Bauer, J. 2005. (Ed.) The Marsh Arabs of Iraq: The legacy of Saddam Hussein and an agenda for restoration and justice. *Carnegie Council Insider,* March/April

Burroughs, E.R. 1976. *Out of time's abyss.* Ace Books.

Churchill, W. 2002. *Struggle for the land: Native American resistance to genocide, ecocide and colonization.* City Lights.

Coles, B. and J. 1989. *People of the wetlands: Bogs, bodies and lake-dwellers.* Thames and Hudson.

Delumeau, J. 2000. *History of paradise: The Garden of Eden in myth & tradition.* Univ. Illinois Press.

Field, H. 1940. *The anthropology of Iraq, No. 2. The lower Euphrates-Tigris region.* Chicago Field Museum.

*Fink, S. 2005. Saving Eden: Can the ecology and the economy of Iraq's once-glorious wetlands be restored? *Discover Magazine* 26(7).

France, R.L. 2004. *Deep immersion: The experience of water.* Green Frigate Books.

France, R.L. (Ed.) 2007. *Healing natures, repairing relationships: New perspectives on restoring ecological spaces and consciousness.* Green Frigate Books.

*France, R.L. (Ed.) 2007. *Sustainable Redevelopment of the Iraqi Marshlands: Lessons and Relevant Applications from around the World.* Routledge.

*France, R.L. 2007. *Back to the garden: Searching for Eden in the Mesopotamian marshes.* Harvard University Press.

*France, R.L. (Ed.). 2007. *Restoring the Iraqi marshlands: Potentials, Perspectives, Practices.* Sussex Academic Press.

Fulanain. 1928. *The marsh Arab.* J.B. Lippincroft Company.

Holland, M.M., P.G. Risser and R.J. Naiman. 1991. *Ecotones: The role of landscape boundaries in the management and restoration of changing environments.* Chapman & Hall.

Holman, T. 1991. *A yearning toward wildness: Environmental quotations from the writings of Henry David Thoreau.* Peachtree.

Jonasson, B. 1995. *Havamal: The sayings of the Vikings.* Gudrun.

Jordan, W.R. III. 2003. *The Sunflower forest: Ecological restoration and the new communion with nature.* Univ. Calif. Press.

Kramer, S.N. 1956. *History begins at Sumer: Thirty-nine firsts in recorded history.* Univ. Penn. Press.

Kramer, S.N. 1969. *Sumerian literary tablets and fragments in the Archeological Museum of Istanbul.* T.T. Kurmu.

Kuper, L. and I. 1982. *Genocide: Its political use in the Twentieth Century.* Yale Univ. Press.

Leoni, E. 2000. *Nostradamus and his prophecies.* Dover.

Maxwell, G. 1966. *People of the reeds.* Pyramid Books.

Mitchell, J.H. 2001. *The wildest place on earth: Italian gardens and the invention of wilderness.* Counterpoint Press.

Nicholson, E. and P. Clark. 2002. *The Iraqi marshlands: A human and environmental study.* Politico's Publishing.

Ochsenschlager, E. L. 1995. Carpets of the Beni-Hassan Village Weavers in Southern Iraq. *Oriental Rug Review* 15(5):12-20.

Ochsenschlager, E. L. 1992. Ethnographic evidence for wood, boats, bitumen and reeds in southern Iraq: Ethnoarchaeology at al-Hiba. *Bulletin on Sumerian Agriculture* 6: 47-78.

Ochsenschlager, E.L. 2004. *Iraq's marsh Arabs in the Garden of Eden.* Univ. Penn. Museum Archeology and Anthropology.

Ochsenschlager, E. L. 1998. Viewing the past: Ethnoarchaeology at al Hiba. *Visual Anthropology* 11 1-2: 103-143.

Patrow, H. 2001. *The Mesopotamian marshlands: Demise of an ecosystem.* UNEP.

Pournelle, J.R. 2003. *Marshland of cities: Deltarc landscapes and the evolution of early Mesopotamian civilization.* Ph. D. Thesis, University of California, San Diego.

Psaki, E.R. and C. Hindley. (Eds.) 2002. *The earthy paradise: The Garden of Eden from antiquity to modernity.* Univ. New York Binghamton.

*Reed, C. 2005. Paradise lost? *Harvard Magazine* 107:3.

Roaf, M. 1990. *Cultural atlas of Mesopotamia and the ancient Near East.* Andromeda Books.

Salim, S.R. 1962. *Marsh dwellers of the Euphrates delta.* The Athlone Press.

Sigrist, M. 1963. *Texts from the British Museum.* CDL Press.

Sigrist, M. 2000. *Sumerian archival texts. III.* Texts from the Yale Babylonian collections. CDL Press.

Simons, G. 1996. *Iraq: From Sumer to Saddam.* St. Martins Press.

Starr, R. F. S. 1939. *Nuzi.* Harvard Univeristy Press.

Thesiger, G. 1964. *The marsh Arabs.* Penguin Books.

Thesiger, Wilfred. 1966. In the marshes of Iraq. *Aramco World*, November-December: 8-19

Thesiger, Wilfred. 2001. *A vanished world.* W. W. Norton.

Uhlein, G. 1983. *Meditations with Hildegard of Bingen.* Bear & Comp.

van De Miercap, M. 1999. *Cuneiform texts and the writing of history.* Routledge.

van Zuylen, G. 1994. *The garden: Visions of paradise.* Thames & Hudson Ltd.

*Wethmann, C. 2005. Mesopotamian marshes and modern development: Conference overview. *Landscape Journal* 24: 223-224.

Wilkinson, T.J. 2003. *Archeological landscapes of the Near East.* Univ. Arizona Press.

Young, Gavin and Nick Wheeler. 1977. The folk that live in the marshes. *Observer Magazine*, May 22: 30-43

Young, G. (photographs by N. Wheeler). 1977. *Return to the marshes: Life with the marsh Arabs of Iraq.* Collins Books.

Young, Gavin and Nick Wheeler. 1976. Water dwellers in a desert world. *National Geographic* 149(4): 502-523.

* Based on conference and related events.

Acknowledgements

This project grew out of background research for a conference I organized in October 2004 investigating the possibility of restoring the ecological and cultural landscapes of the Mesopotamian marshes in southern Iraq. The conference and related events were hosted and sponsored by the Design School, the Kennedy School of Government, and the Semitic and Peabody Museums of Harvard University, and the Carnegie Council for Ethics and International Affairs in association with New York University's Environmental Conservation Education Program. Corporate sponsors for the conference and production of this book included Applied Ecological Services, the Canadian International Development Agency, CH2M HILL, Design Workshop, Dharma Living Systems, Ducks Unlimited, the Iraq Foundation Eden Again Project, Jones & Jones, Michael Baker Corporation, North American Wetland Engineering, the United States Development Agency, as well Banrock Station Winery and the Cambridge Brewing Company. The major sponsor for production of this book was the William F. Milton Fund of Harvard University.

I would like to thank Pat Kervick and India Spartz of the Peabody Museum at Harvard University and Nina Cummings of the Chicago's Field Museum for use of the wonderful archival photographs of the marsh dwellers. I would especially like to thank James Armstrong for many discussions concerning, and help with selecting and writing about, the artifacts from the Semitic Museum at Harvard University that were ably photographed by Doug Cogger from the Design School. Additional conversations with Joseph Greene from the Semitic Museum did much to help orient me in the complicated world of the ancient Near East. Natalie DeNormandie is thanked for the compelling initial graphic designs that were pushed through book production ably by Jennifer Brown of Opaque

Design & Print Production, as is Doug Cogger for photographing the museum replicas and artifacts (including the cover).

Finally, I would like to especially thank my contributing essayists, Baroness.Emma Nicholson, Nik Wheeler and Rasheed Bander Al-Khayoun, whose informed and powerful words do much to bring readers from the ancient to the recent past in these pages. These essays were derived from transcriptions made of presentations given at the October 2004 converence and any errors reflect my own editing.

ROBERT LAWRENCE FRANCE is an ecologist at the Harvard Design School where he teaches courses on water management, restoration design, and environmental theory. Dr. France organized the recent Harvard University conference on restoring the ecological and cultural landscapes of the Mesopotamian marshes in southern Iraq. He has authored close to two hundred technical papers and is the author or editor of numerous books including *Deep Immersion: The Experience of Water* and *Profitably Soaked: Thoreau's Engagement with Water*, both published by Green Frigate Books. Dr. France will also be the author or editor of three forthcoming books derived from the Harvard conference: the two technical publications, *Sustainable Development of the Iraqi Marshlands: Lessons and Relevant Applications From Around the World*, and *Restoring the Iraqi Marshlands: Potentials, Perspectives, Practices*, as well as the popular title, *Back to the Garden: Searching for Eden in the Mesopotamian Marshes*.

EDWARD L. OCHENSCHLAGER is Professor Emeritus at Brooklyn College where he first served as chairman of the Classics Department and later as chairman of the Department of Anthropology and Archaeology. He was director of Excavations at Thmuis and Taposiris Magna in Egypt, American Field Director and Principal Investigator of the Smithsonian-Archaeological Institute of Belgrade Excavations at Sirmium in Yugoslavia. He served as Assistant Field Director of the Institute of Fine Art's of New York University and the Brooklyn Museum's Mendes Excavation, of the Institute of Fine Art's and the Metropolitan Museum of Art's Excavations at Al-Hiba in Iraq, and most recently was Assistant Director of the Institute of Fine Art's Excavations at Shibam in Yemen. He is author of *Iraq's Marsh Arabs in the Garden of Eden*, dealing with his ethno-archaeological research at al-Hiba, and was a speaker at Harvard University in an event associated with the 2004-2005 exhibition of early marsh Arab photographs.

NIK WHEELER began his photographic career as a combat photographer for United Press International in Vietnam. In the 1970s he freelanced out of Beirut covering major news events such as the Jordan Civil War and the 1973 Middle East War for *Time* and *Newsweek*. It was during this period that he first visited Iraq to document the Marsh Arabs for *National Geographic* and subsequently produced his first book, with text by Gavin Young, *Return to the Marshes*. A second book on Iraq, *Land of Two Rivers*, was published in 1980. After three years in Paris where he worked for Sipa Press, he moved to Los Angeles and concentrated on travel photography, doing assignments for major travel magazines. In 1988 he was named "Travel Photographer of the Year" by SATW. His travel photos have appeared on more than two hundred magazine covers (including the January 2005 issue of the *Harvard Magazine*) and he has illustrated more than twenty books. He is currently working on a coffee table book entitled *Island Dreams–Caribbean*, and was a participant at the Harvard conference on the Iraqi marshlands from where this contribution for publication was adapted.

BARONESS EMMA NICHOLSON of Winterbourne was elected as a Member of the European Parliament for the South East region of England in June 1999 and re-elected as MEP for the same region in June 2004. She served as Vice-Chairman of the Conservative Party from 1983-1987, as Member of the House of Commons from 1987-1997, and was created a Life Peer in 1997. Her NGO work includes the chairmanship of the AMAR International Charitable Foundation, Presidency of the Caine Prize for African Writing, Trusteeship of the Booker Prize for English fiction, and of the Booker Prize for Russian fiction. She serves as World Health Organization Envoy for Health, Peace and Development. Baroness Nicholson's healthcare work in developing countries has been focused on preventative care, public health provision, capacity building, and sustainable development. She has worked on projects in southern and western Africa, Lebanon, Romania, Iran, Afghanistan, in addition to Iraq. Baroness Nicholson's work on education has been closely tied to UNESCO. She has also initiated a high level, continuing dialogue and conference on Islam and other civilizations. Baroness Nicholson is the Co-Editor of *The Iraqi Marshlands: A Human and Environmental Study*, and was a participant at the Harvard conference on the Iraqi marshlands from where this contribution for publication was adapted.

RASHEED B. AL-KHAYOUN was born in the marshes of Iraq. He left home in 1979 due to political reasons and immigrated to Bulgaria and then Aden, Yemen. During his time away from Iraq he completed his MA and PhD studies in Islamic philosophy at Sofia University, Bulgaria. He has published hundreds of articles and conducted research on many aspects of the Islamic faith and also on the Al-Ahwar, or marshlands of southern Iraq. Throughout his career as a researcher he has published a number of books on subjects concerning religion in Iraq and Islamic philosophy and history, including *Religions and Sects in Iraq*, and *Mutazilat Basra and Baghdad*. Dr. Al-Khayoun has worked as Cultural Editor for *Al-Mutamer* newspaper and at present works as a researcher for Al-Hura TV and as a writer for *Al-Sharig Al-Awsat* newspaper. He was a participant at the Harvard conference on the Iraqi marshlands from where this contribution for publication was adapted.

Also by Robert Lawrence France

Authored

*Deep Immersion: The Experience of Water**
Wetland Design: Principals and Practices for Landscape Architects and Land-Use Planners
Introduction to Watershed Development: Understanding and Managing the Impacts of Sprawl
Aquatic Responses to Watershed Clearcutting: Implications for Forestry and Fisheries Management (forthcoming)
Back to the Garden: Searching for Eden in the Mesopotamian Marshes (forthcoming)
Singular Element, Interestingly Strange: Thoreau's Natural History of Water (forthcoming)
Sustainable Development of the Iraqi Marshlands: Lessons and Relevant Applications From Around the World (forthcoming)
Restoring the Iraqi Marshlands: Potentials, Perspectives, Practices (forthcoming)

Co-authored

Road Ecology: Science and Solutions
Manganese in the Canadian Environment

Edited

*Profitably Soaked: Thoreau's Engagement With Water**
Handbook of Water Sensitive Planning and Design
Reflecting Heaven: Thoreau on Water
Facilitating Watershed Management: Fostering Awareness and Stewardship
*Ultreia! Onward! Progress of the Pilgrim**
*Healing Natures, Repairing Relationships: New Perspectives on Restoring Ecological Spaces and Consciousness**
Handbook of Landscape Regenerative Development and Design (forthcoming)

* Published by Green Frigate Books

Other books by Green Frigate Books

GREEN FRIGATE BOOKS

"THERE IS NO FRIGATE LIKE A BOOK"

Words on the page have the power to transport us, and in the process, transform us. Such journeys can be far reaching, traversing the landscapes of the external world and that within, as well as the timeapes of the past, present and future.

Green Frigate Books is a small publishing house offering a vehicle—a ship—for those seeking to conceptually sail and explore the horizons of the natural and built environments, and the relations of humans within them. Our goal is to reach an educated lay readership by producing works that fall in the cracks between those offered by traditional academic and popular presses.

Wetlands of Mass Destruction